Coherent States in Gauge Theories and Applications in Collider Physics

Coherent States in Gauge Theories and Applications in Collider Physics

Editor

Mario Greco

INFN & Università Degli Studi Roma Tre, Italy

World Scientific

NEW JERSEY • LONDON • SINGAPORE • BEIJING • SHANGHAI • HONG KONG • TAIPEI • CHENNAI • TOKYO

Published by

World Scientific Publishing Co. Pte. Ltd.
5 Toh Tuck Link, Singapore 596224
USA office: 27 Warren Street, Suite 401-402, Hackensack, NJ 07601
UK office: 57 Shelton Street, Covent Garden, London WC2H 9HE

British Library Cataloguing-in-Publication Data
A catalogue record for this book is available from the British Library.

The author and publisher would like to thank the following publishers of the various journals for their assistance and permission to include the selected reprints found in this book:

American Physical Society (*Phys. Rev. Lett.*); Elsevier (*Nucl. Phys. B, Phys. Lett. B.*); Società Italiana di Fisica (*Il Nuovo Cimento*)

COHERENT STATES IN GAUGE THEORIES AND APPLICATIONS IN COLLIDER PHYSICS

ISBN 978-981-121-389-2

For any available supplementary material, please visit
https://www.worldscientific.com/worldscibooks/10.1142/11654#t=suppl

To my wife, Halina.

Preface

The idea of coherent states was suggested in QED in the middle of 1960's and in QCD at the end of the 1970's to introduce a realistic definition of initial and final states, where the number of quanta is not determined, unlike the usual approach in perturbation theory. In addition to simply solving conceptual infrared divergences of the S-Matrix, it allows a description of the properties of the QED and QCD radiation to all orders in perturbation theory.

In lepton colliders, it gives a precise determination of the line-shape production of the J/ψ, the Z and the Higgs bosons, while in QCD it allows a realistic description of quark and gluon jets, of the transverse momentum distribution of W, Z and H bosons produced in hadron collisions, and other related quantities.

The book consists of a collection of the articles published by the author and his collaborators in over fifty years, with an introduction highlighting the main results of each article and the various links between them.

This book is dedicated to Bruno Touschek who inspired my first work on coherent states, conceived and realised AdA, the first e^+e^- collider, opening the new era of the collider physics.

Contents

Introduction

The notion of "Coherent States" was first introduced by *R. Glauber* [1] in the field of quantum optics. Later, the term "coherent states" was extended in its usage to broader areas of physics such as quantum mechanics, condensed matter physics, mathematical physics, quantum field theories and others. A vast collection of original papers and references in which the concept of coherent states plays a significant role can be found in the book "*Coherent States*" by *Klauder and Skagerstam* [2].

In gauge theories the notion of coherent states was associated to the study of the infrared behaviour of the theory, starting from QED in particular. Since the basic work of *Bloch and Nordsieck* [3], which was refined and improved over the years [4–7], the standard approach involves the computation of the perturbative amplitudes which are formally singular, but eventually the perturbative divergences vanish in the physical cross sections after summing up the real and virtual photon contributions. The alternative approach, based on the idea of coherent states, is to directly formulate an infrared finite S-matrix by choosing appropriate asymptotic states, which are not eigenstates of the photon number operator but possess the form of coherent states. This approach started from the works of *Chung* [8], independently of *Greco and Rossi* [9], and was further developed by a number of authors, e.g. [10–12]. For a review see also Ref. [13]. The approach was partially extended to the more complicated case of non-Abelian gauge theories, particularly QCD, [15–22] and more recently to perturbative quantum gravity [23].

One of the great merits of this approach is to provide in a straightforward manner the final result of a computed physical observable in exponentiated form which sums up to all orders of the leading and sometimes also the next-to-leading contributions of the perturbation theory. This is very convenient when the normal calculation of the lowest orders does not provide the right or complete answer to the physical problem. Of course additional finite terms must be usually added in order to provide the accuracy required in the theoretical result. The physics explored with both lepton and hadron colliders has provided in recent years many examples of such situations. The collective effect of the soft photons emitted in e^+e^- annihilation in the case of a narrow resonance production, as the J/ψ or the Z boson, provides the clearest example in QED. Similarly the calculation of the transverse momentum distribution of a W or Z or Higgs bosons produced inclusively in hadron-hadron collisions is another typical example in QCD.

This book presents a collection of articles published by the author and his collaborators in many years, starting from the introduction of coherent states in QED in the sixties and in QCD in the late seventies, to their development in the phenomenology of the Standard Model, with a particular emphasis on the applications of these ideas in the physics of both lepton and hadron colliders, including the discussion of the current problems in Higgs physics at the future electron and muon colliders. In the following, we will briefly report on the various articles, where our attempts highlight the main results of each paper and point out the relative links among them.

In the first QED paper "*A Note on the Infrared Divergence*" [9] a finite S-matrix element **M** is introduced by defining new realistic final states which contain an undetermined number of soft photons in addition to the detected charged particles. Then **M** is shown to be directly comparable to the observable cross section which results to be proportional to $|\mathbf{M}|^2$, and separable as $\mathbf{M} = \mathrm{A}\,\mathbf{M}'$, where A depends on the cut-off for soft photons and $\mathbf{M}'$ is finite. This is achieved by introducing the notion of the "classical currents" associated to the initial and final charged particles:

$$j_\mu^{(\mathrm{i,f})}(k) = \frac{ie}{(2\pi)^{3/2}} \sum_{1}^{(\mathrm{i,f})} \varepsilon_1 \frac{p_\mu^{(1)}}{(p_1 \cdot k)},$$

which are the sources of the quantized electromagnetic field associated to the emission of the soft photons.

The discovery of the very narrow J/ψ particle in 1974 in the electron-positron annihilation motivated *Greco, Pancheri and Srivastava* [24] to extend the above concept to the following one:

$$j_\mu^{\mathrm{R}}(k) = \frac{W - M + \frac{1}{2}i\Gamma}{W - M - k + \frac{1}{2}i\Gamma} j_\mu^{(\mathrm{i})}(k) + j_\mu^{(\mathrm{f})}(k),$$

where M and Γ are the mass and width of the resonance **R** produced in the e^+e^- annihilation channel with total energy W. This modification of the current relative to the initial state takes into account the finiteness of the time interval between the formation of the resonance in the annihilation process and the creation of the final state, as it can be seen from the Fourier transform of the above equation. Then the main radiative correction factor to the Born cross section becomes $\approx (\Gamma/\mathrm{M})\, \exp[(4\alpha/\pi)\ln(\mathrm{W}/\mathrm{m}_e)]$, where m_e is the electron mass. Physically, this is understood as follows: the width of the resonance Γ provides a natural cut-off in damping the energy loss in the initial state. The detailed analysis of the process led [24] to the first and complete description of the radiative effects in the J/ψ production in e^+e^- annihilation, including the interference effects.

The extension of the above result to the case of Z boson production in the framework of the Weinberg–Salam Standard Model followed in 1980 [25], with the soft photon effects resummed to all orders, and no restriction on the relative magnitudes of the experimental energy resolution $\Delta\omega$ and the Z-width Γ. This paper was at the base of many phenomenological analyses of the experimental data at LEP and SLC.

In the middle of the 1980's a new approach was successfully proposed by *Kuraev and Fadin* [26] for studying the radiative corrections to e^+e^- annihilation at high energies to all leading orders, and generalized by other authors to higher accuracy [27]. It is based on the idea of the electron and positron structure functions and is very useful indeed for the numerical simulations of radiation processes as well as for the analysis of the data. Then in Ref. [28] the methods of coherent states and of structure functions were considered in detail for a precise evaluation of the radiative effects at LEP/SLC. They were explicitly shown to give identical results for the exponentiated infrared factors and the $o(\alpha)$ terms corresponding to the initial and final state radiation. The combined use of both approaches was shown to offer two complementary methods for improving the theoretical predictions to an accuracy better than 1%, including interference effects, and provide simple analytical formulae of immediate phenomenological application.

The Higgs boson discovery at the LHC in 2012 [29, 30] has opened a new era of particle physics. It is the first elementary scalar particle ever observed in Nature; hence, its properties need to be thoroughly scrutinized. Future lepton collider Higgs factories [31–36] have been proposed to study the Higgs boson properties to great accuracies because of the much more favourable experimental environment than that at hadron colliders. Amongst many candidates of Higgs factories, the possibility of s-channel resonant production is especially important. The muon collider Higgs factory could produce the Higgs boson in the s-channel and perform an energy scan to map out the Higgs line-shape at tens of MeV level. This approach would provide the most direct measurement of the Higgs boson's total width and the Yukawa coupling to muons. Also the possibility of an ultra high luminosity electron-positron collider for the Higgs resonant production has been proposed [37], providing a possible opportunity to observe the Higgs signal and thus the determination of the Yukawa coupling to electrons.

Then in Ref. [38], following the previous analyses for the J/ψ and Z production, we have shown immediately that in the case of a muon collider the very important corrections due to Initial State Radiation (ISR) drastically change the lowest order Higgs production results, and must be taken in full account for an accurate description of the Higgs total and partial widths. In addition, due to the narrow width of the Higgs boson, of about 4 MeV as predicted by the Standard Model, it would be extremely demanding for the collider energy resolution to reach a similar value in order to adequately study the physical width. Then by convoluting the Breit–Wigner resonance for the Higgs signal and the Gaussian distribution for the profile of beam energy spread — including the ISR — in a more complete analysis *Greco, Han and Liu* [39] have illustrated the combined impact on a muon collider for the experimental study of the Higgs line-shape as well as for the machine design of the beam energy spread. Of course those effects would be notably stronger for the future e^+e^- colliders in the case of s-channel Higgs production because of the lighter electron mass. Alternatively, in observing the Higgs signal mainly through the associated HZ production, e.g. through the reaction $e^+e^- \to ZH$ at different c.m. energies, the ISR effects are not so strong but could, in principle, be sizable as shown in the detailed analysis of Ref. [40].

The first extension of the concept of coherent states in QCD appeared in 1978, with the paper "*Coherent State Approach to the Infrared Behaviour of Non-Abelian Gauge Theories*" by *Greco, Palumbo, Pancheri and Srivastava* [13]. There, the infrared behaviour of the matrix elements between coherent states of definite colour was investigated in non-Abelian gauge theories. The matrix elements were shown to be finite to the lowest nontrivial order, and to all orders if the soft gluon formula holds to all orders, at the leading logarithmic approximation. This was achieved by extending the concept of classical currents associated to the external particles to incorporate both the new properties of colour and the appearance of the effective coupling constant. Then *Curci and Greco* further extended in Ref. [41] the formalism of coherent states to the problem of mass singularities, by including the hard radiation emitted within an arbitrarily small angle but with fixed value δ. The obvious motivation for this analysis was the very close connection between hard processes and mass singularities in QCD. Such an extension provided one with new matrix elements free of mass singularities at all orders of perturbation theory, in the leading logarithmic approximation, in agreement with the *Kinoshita–Lee–Nauenberg* theorem [5, 6]. As a nice feature of the formalism the contribution of hard and collinear quanta emitted from any external leg of a diagram was easily obtained in exponentiated form. Those results provided a very simple tool for evaluating, to all orders, the finite corrections to hard processes. In addition, because the usual measuring procedure of the charge of a particle fails in the limit $\mathrm{m} \to 0$, one is naturally led to introduce an effective coupling constant in the definition of the classical current which appears in the coherent states. As a consequence of this approach *Curci and Greco* in Ref. [42] had examined the effect of soft gluon emission in hard processes to all orders in QCD. Large corrections had been found, which sensibly modified the parton model results, and showed the relevance of the appropriate kinematical limits in lepto-production and in the Drell–Yan process. The formalism of coherent states was then used by *Curci, Greco and Srivastava* in Refs. [43, 44] to study the jet structure in QCD. Simple formulae were obtained for a general jet cross section in terms of an infinite number of massless quarks and gluons.

In addition, the first derivation was given in QCD for the probability distribution of the jet transverse momentum which included the correlations induced by the momentum conservation:

$$\frac{\mathrm{d}P}{\mathrm{d}K_\perp} = K_\perp \int_0^\infty \xi \mathrm{d}\xi J_0(\xi K_\perp) \exp\Big(- \frac{16}{3\pi} \int_0^{k_\perp} \frac{\mathrm{d}q_\perp}{q_\perp} \ln(Q/q_\perp)\alpha_s(q_\perp)[1 - J_0(\xi q_\perp)]\Big),$$

where $\alpha_s(q_\perp)$ is the running coupling constant.

This formula, which also describes the transverse momentum distributions of a Drell–Yan pair to all orders in the leading logarithmic approximation, or a W/Z produced in hadronic collisions, was the completely correct formula for the leading order Sudakov factor, after the wrong result given in a previous paper by *Dokshitzer, Diakonov and Troyan* [45].

Phenomenologically, it was successfully compared with the experimental data taken by the PLUTO collaboration at the $\mathrm{e^+e^-}$ storage rings, DORIS and PETRA, in the analysis performed in Ref. [46] in collaboration with the present author.

In those years the Rome theory group contributed very much to the development of perturbative QCD, in particular with the calculation of next-to-leading corrections to Drell–Yan processes by *Altarelli, K. Ellis and Martinelli* [47]. Then the important problem was that of realizing a smooth match between the perturbative and the Sudakov components. When the data on the W and Z production at CERN from UA1 and UA2 Collaborations became available, the right theoretical prediction was finally given in the paper by *Altarelli, K. Ellis, Greco and Martinelli* [48]. That was an important result because it essentially contained all the crucial ingredients that describe the physics of this phenomenon. In the subsequent years the accuracy has been much improved with the computation of subleading effects and with several other refinements, but the essential points were all present in that paper, where the accuracy was adequate for the quality of the first data. The same techniques are at present applied to the calculation of the p_T distribution of the Higgs boson produced by gluon fusion (see, for example, Ref. [49]).

This concludes the introduction to the motivations and the impact of the results presented in the various articles on the physics of the lepton and hadron colliders. The original papers will follow in the subsequent sections.

References

1. R. J. Glauber, "*Quantum theory of optical coherence*", Phys. Rev. 130 (1963) 2529; "*Quantum Optics and Electronics*", eds. C. DeWitt, A. Blandin and C. Cohen-Tannoudji, Gordon and Breach, New York, 1964.
2. "*Coherent States*", eds. John R. Klauder and Bo-Sture Skagerstan, Word Scientific, 1985.
3. F. Bloch and A. Nordsieck, "*Note on the Radiation Field of the electron*", Phys. Rev. 52 (1937) 54.
4. D. R. Yennie, S. C. Frautschi and H. Suura, "*The infrared divergence phenomena and high-energy processes*", Annals Phys. 13 (1961) 379.
5. T. Kinoshita, "*Mass singularities of Feynman amplitudes*", J. Math. Phys. 3 (1962) 650.
6. T. D. Lee and M. Nauenberg, "*Degenerate Systems and Mass Singularities*", Phys. Rev. 133 (1964) B1549.
7. S. Weinberg, "*Infrared photons and gravitons*", Phys. Rev. 140 (1965) B516.
8. V. Chung, "*Infrared Divergence in Quantum Electrodynamics*", Phys. Rev. 140 (1965) B1110.
9. M. Greco and G. Rossi, "*A Note on the Infrared Divergence*", Il Nuovo Cimento 50 (1967) 168.
10. T. W. B. Kibble, "*Coherent Soft-Photon States and Infrared Divergences. I. Classical Currents*", J. Math. Phys. 9 (1968) 315; "*Coherent soft-photon states and infrared divergences. ii. mass-shell singularities of Green's functions*", Phys. Rev. 173 (1968) 1527.
11. D. Zwanziger, "*Reduction formulas for charged particles and coherent states in quantum electrodynamics*", Phys. Rev. D7 (1973) 1082.
12. P. P. Kulish and L. D. Faddeev, "*Asymptotic conditions and infrared divergences in quantum electrodynamics*", Theor. Math. Phys. 4 (1970) 745.
13. N. Papanicolaou, "*Infrared Problems in Quantum Electrodynamics*", Phys. Rept. 24, (1976) 229.

14. M. Greco, F. Palumbo, G. Pancheri-Srivastava and Y. Srivastava, "*Coherent state approach to the infra-red behaviour of non-abelian gauge theories*", Phys. Lett. B77 (1978) 282.
15. D. R. Butler and C. A. Nelson, "*Non-abelian structure of yang-mills theory and infrared-finite asymptotic states*", Phys. Rev. D18 (1978) 1196.
16. C. A. Nelson, *Origin of Cancellation of Infrared Divergences in Coherent State Approach: Forward Process* $q\ q \rightarrow q\ q$ *gluon*", Nucl. Phys. B181 (1981) 141.
17. H. D. Dahmen and F. Steiner, "*Asymptotic Dynamics of QCD, Coherent States and the Quark Form-factor*", Z. Phys. C11 (1981) 247.
18. J. Frenkel, J. Gatheral and J. Taylor, "*Asymptotic states and infrared divergences in non-abelian theories*", Nucl. Phys. B194 (1982) 172.
19. S. Catani and M. Ciafaloni, "*Generalized coherent state for soft gluon emission*", Nucl. Phys. B249 (1985) 301.
20. S. Catani, M. Ciafaloni and G. Marchesini, "*Non-cancelling infrared divergences in QCD coherent states*", Nucl. Phys. B264 (1986) 588.
21. S. Catani and M. Ciafaloni, "*Gauge covariance of QCD coherent states*", Nucl. Phys. B289 (1987) 535.
22. M. Ciafaloni, "*The QCD coherent state from asymptotic dynamics*", Phys. Lett. B150 (1985) 379.
23. J. Ware, R. Saotome and R. Akhoury, "*Construction of an asymptotic S matrix for perturbative quantum gravity*", J HEP10 (2013) 159.
24. M. Greco, G. Pancheri and Y. Srivastava, "*Radiative Corrections for Colliding Beam Resonances*", Nucl. Phys. B101 (1975) 234.
25. M. Greco, G. Pancheri and Y. Srivastava, "*Radiative Corrections to* $e^+e^- \rightarrow \mu^+\mu^-$ *around the* Z_0", Nucl. Phys. B171 (1980) 118; Erratum: Nucl. Phys. B197 (1982) 543.
26. E. A. Kuraev and V. S. Fadin, "*On Radiative Corrections to* e^+e^- *Single Photon Annihilation at High Energy*", Sov. J. Nucl. Phys. 41 (1985) 466.
27. O. Nicrosini and L. Trentadue, "*Soft photons and second order radiative corrections to* $e^+e^- \rightarrow Z_0$", Phys. Lett. B196 (1987) 553.
28. F. Aversa and M. Greco, "*Coherent States and Structure Functions in QED*", Phys. Lett. B228 (1989) 134.
29. ATLAS Collaboration, G. Aad *et al.*, "*Observation of a new particle in the search for the Standard Model Higgs boson with the ATLAS detector at the LHC*", Phys. Lett. B716 (2012) 1.
30. CMS Collaboration, S. Chatrchyan *et al.*, "*Observation of a new boson at a mass of 125 GeV with the CMS experiment at the LHC*", Phys. Lett. B716 (2012) 30.
31. V. D. Barger, M. S. Berger, J. F. Gunion and T. Han, "*Higgs Boson physics in the s channel at* $\mu^+\mu^-$ *colliders*", Phys. Rept. 286 (1997) 1.
32. Y. Alexahin *et al.*, "*The Case for a Muon Collider Higgs Factory*", Snowmass on the Mississippi, Minneapolis, MN, USA, July 29–August 6, 2013. arXiv:1307.6129.
33. H. Baer *et al.*, "*The International Linear Collider Technical Design Report*", Vol. 2: Physics, arXiv:1306.6352.
34. C. Rubbia, "*A complete demonstrator of a muon cooled Higgs factory*", arXiv:1308.0612.
35. TLEP Design Study Working Group Collaboration, M. Bicer *et al.*, "*First Look at the Physics Case of TLEP*", JHEP 01 (2014) 164.
36. M. Ahmad *et al.*, "*CEPC-SPPC Preliminary Conceptual Design Report. 1: Physics and Detector*", (2015).
37. A. Abada *et al.*, "*FCC Physics opportunities*", Eur. Phys. J. C (2019) 79–474.
38. M. Greco, "*On the study of the Higgs properties at a muon collider*", Mod. Phys. Lett. A 30 (2015) 1530031.

39. M. Greco, T. Han and Z. Liu, "*ISR effects for resonant Higgs production at future lepton colliders*", Phys. Lett. B 763 (2016) 409.
40. M. Greco, G. Montagna, O. Nicrosini, F. Piccinini and G. Volpi, "*ISR corrections to associated H Z production at future Higgs factories*", Phys. Lett. B 777 (2018) 294.
41. G. Curci and M. Greco, "*Mass Singularities and Coherent States in Gauge Theories*", Phys. Lett. 79B (1978) 406.
42. G. Curci and M. Greco, "*Large Infrared Corrections in QCD Processes*", Phys. Lett. 92B (1980) 175.
43. G. Curci, M. Greco and Y. Srivastava, "*Coherent Quark-Gluon Jets*", Phys. Rev. Lett. 43 (1979) 834.
44. G. Curci, M. Greco and Y. Srivastava, "*(QCD) Jets from Coherent States*", Nucl. Phys. B159 (1979) 451.
45. Yu. Dokshitzer, D. Diakonov and S. Troyan, "*Hard Semi-inclusive Processes in QCD*", Phys. Lett. B78 (1978) 290.
46. PLUTO Collaboration, Ch. Berger *et al.*, "*Observation of QCD Effects in Transverse Momenta of e^+e^- Jets*", Phys. Lett. 100B (1981) 351.
47. G. Altarelli, R. K. Ellis and G. Martinelli, "*Large Perturbative Corrections to the Drell–Yan Process in QCD*", Nucl. Phys. B157 (1979) 461.
48. G. Altarelli, R. K. Ellis, M. Greco and G. Martinelli, "*Vector Boson Production at Colliders: a Theoretical Reappraisal*", Nucl. Phys. B246 (1984) 12.
49. G. Bozzi, S. Catani, D. de Florian and M. Grazzini, "*Higgs boson production at the LHC: Transverse-momentum resummation and rapidity dependence*", Nucl. Phys. B791 (2008) 1.

IL NUOVO CIMENTO VOL. L A, N. 1 1° Luglio 1967

A Note on the Infra-Red Divergence.

M. GRECO and G. ROSSI
Laboratori Nazionali del CNEN - Frascati

(ricevuto il 24 Gennaio 1967)

Summary. — In this paper we show how the infra-red divergence can be eliminated in the matrix element, provided that physically true final states for an experiment involving creation and destruction of charged particles are used.

1. – Introduction.

It is well known that no physical process involving the creation and destruction of charged particles can take place without the emission of soft photons. This fact together with the observation that the number of soft photons emitted in a reaction can never really be measured, leads one to introduce a new definition of the final states. This definition is closer to physical reality than the use of states which are diagonal in the number of photons.

We shall show that with this redefinition of the final state it is possible to determine a matrix element M for the process, which is

1) finite and does not therefore present an infra-red divergence,

2) directly comparable to the experimental cross-section which results to be proportional to $|M|^2$,

3) and separable: $M = AM'$, where A depends on the cut-off for soft photons and M' is finite.

The present method therefore leads to a compensation of the infra-red divergence for the matrix element itself and does not require—as in the usual method—the compensation of the terms corresponding to real and virtual photons in the expression for the cross-section. This simplification is due to the fact that the Bloch-Nordsieck theorem allows one to predict the phases of the emitted photons.

4140

The new final states are not eigenstates of the energy: this corresponds exactly to the experimental situation, which does not allow a check of energy conservation to any desired accuracy.

2. – Relativistic deformation of final states.

We consider a process

$$\mathrm{A}+\mathrm{B}\to\mathrm{C}+\mathrm{D}+\dots, \tag{1}$$

in which at least part of the particles A, B, C, D, ... involved are charged. What one actually observes is the process

$$\mathrm{A}+\mathrm{B}\to\mathrm{C}+\mathrm{D}+\dots+\Gamma, \tag{2}$$

where Γ stands for any number of photons, limited only by the condition that the total 4-momentum of the emitted photons should be contained in that part of the forward light cone, which is defined by the energy and momentum resolution of the experimental apparatus. In the following we shall assume that there is energy resolution only—and no momentum resolution— and we assume that $\Delta\omega \ll E$, where for example E is the centre-of-mass energy of an incident particle.

The final state of reaction (1) will be called $|f)$. For describing reaction (2) we introduce a state vector $|f')$ defined as

$$|f') = e^{iA_c}|f) \tag{3}$$

with

$$A_c = \int \mathrm{d}^4x\, j_\mu(x)\, A_\mu(x)\,, \tag{4}$$

where $A_\mu(x)$ is the 4-potential of the electromagnetic field in the interaction representation and $j_\mu(x)$ is a c-number current defined as

$$j_\mu(x) = (2\pi)^{-4}\int \mathrm{d}^4k\, j_\mu(k)e^{ikx} \tag{5}$$

and

$$j_\mu(k) = ie(2\pi)^{-3}\sum_i \varepsilon_i P_{i,\mu}(P_i k)^{-1} \qquad \text{for } K_0 < \Delta\omega \text{ and zero otherwise}\,. \tag{6}$$

The signature ε_i is $+1$ for positive outgoing and negative incoming particles and -1 otherwise and p_i with $p_{i,0}>0$ is the energy-momentum vector of the i-th particle. We use the metrics $g_{00}=1$, $g_{11}=g_{22}=g_{33}=-1$. The current defined by eq. (6) is the Fourier component of the classical current accompanying the destruction of the particles A, B and the creation of the particles C, D... The cut-off introduced in this equation is not completely realistic, since it concerns the energy of the single photons, rather than the total energy carried away by the electromagnetic field. We shall discuss this point in Sect. 5.

The operator e^{iA_c} defined in eq. (3) is unitary. The new states $|f')$ are therefore normalized provided that the old states $|f)$ were normalized. In the new frame of reference the matrix element for the transition $i\to f$ is given by

$$\bar{M}=(f'|S|i)=(f|e^{-iA_c}T(e^{iL})|i)\,; \tag{7}$$

L is the action integral describing the interaction that leads to process (1).

The matrix element $\bar{M}$ can be formally written as

$$\bar{M}=\langle f|(e^{-iA_c}-\mathbf{1})S|i\rangle+\langle f|S|i\rangle=M_c+M_{el}\,. \tag{8}$$

Of course such a definition has no physical meaning; on the other hand mathematically M_c and M_{el} are clearly defined by eq. (8) and serve as means to compare the present formalism with perturbation theory.

We emphasize that M is formally the matrix-element of the operator $S'=\exp[-iA_c]S$ between the states $\langle f|$ and $|i\rangle$ which do not include outgoing soft photons: in other words the infinite number of soft photons created by S in the final state is destroyed by $\exp[-iA_c]$.

3. – Finiteness of $\bar{M}$.

We are now going to show that $\bar{M}$ is finite, *i.e.* that, apart from ultra-violet divergences, it does not present any infra-red divergence. More precisely we shall prove that the expression

$$|\bar{M}|^2=|M_c|^2+|M_{el}|^2+2\,\mathrm{Re}\,(M_c M^*_{el})$$

is finite.

For this purpose we shall use the well-known theorem [1,2] which states that the infra-red divergences in the cross-section concerning the bremsstrahlung graphs are exactly compensated to any order in the fine-structure constant, by analogous divergences in the corresponding graphs with virtual photons. Then, as

$$M_{el}=\langle f|S|i\rangle\,,$$

[1] J. M. Jauch and F. Rohrlich: *Helv. Phys. Acta*, **27**, 613 (1954).

[2] J. M. Jauch and F. Rohrlich: *The Theory of Electrons and Photons* (New York, 1955).

it will be sufficient to show that, in the limit $\omega_j \to 0$ (ω_j being the energy of of the j-th bremsstrahlung photon), the cross-section for the m-th order bremsstrahlung calculated with the usual methods of electrodynamics, coincides with the equal-order term of

$$|M_r|^2 + 2\,\mathrm{Re}\,(M_r M_{el}^*)\,.$$

From eq. (7), with our definitions, we have

$$(9)\qquad \sum_{n+t-s-2m=\text{const}} \frac{(-i)^t}{t!} A_c^t \frac{(i)^n}{n!} T(L^n) = \\ = \frac{1}{m!}\left\{\frac{1}{2}\int \mathrm{d}^4k\,\delta(k^2)\theta(k_0) j_\mu(k) j_\mu^*(k)\right\}^m \langle f|\frac{(i)^s}{s!} T(L^s)|i\rangle\,.$$

Now we can easily find the term of order α^{m+s} in $|M_r|^2 + 2\,\mathrm{Re}\,(M_r M_{el}^*)$:

$$(10)\qquad [|M_r|^2 + 2\,\mathrm{Re}\,(M_r M_{el}^*)]_{(m+s)} = \frac{1}{m!}\left\{\int \frac{\mathrm{d}^3k}{2k_0} j_\mu(k) j_\mu^*(k)\right\}^m \left|\langle f|\frac{(i)^s}{s!} T(L^s)|i\rangle\right|^2\,.$$

The definition of $j_\mu(k)$ (eqs. (5) and (6)) sets the upper limits on the integrals which appear in eq. (10).

The analogous term calculated with the usual final states is

$$(11)\qquad \frac{1}{m!}\left\{\int \frac{\mathrm{d}^3k}{2k_0} j'_\mu(k) j'^*_\mu(k)\right\}^m \left|\langle f|\frac{(i)^s}{s!} T(L^s)|i\rangle\right|^2\,,$$

where now, adopting the notation already introduced, $j'_\mu(k)$ is defined by

$$(12)\qquad j'_\mu(k) = ie(2\pi)^{-\frac{3}{2}} \sum_i \varepsilon_i P_{i\mu}(P_i k)^{-1}$$

without any restriction imposed on the values which k can assume. The upper limits in the integrals are fixed by the global condition

$$\sum \omega_k \leqslant \Delta\omega\,.$$

Although eqs. (10) and (11) would not seem to be exactly equal, we must recall that the description of the final states so far given in our formalism, is not completely realistic (compare Sect. **2**) and therefore actually we ought to cut also the integrals in (10) according to the above global condition.

As we shall discuss later (Sect. **5**) the difference between eqs. (10) and (11) is given only by a normalization factor very near to 1.

In this way we have shown the finiteness of $|\overline{M}|^2$ and consequently of $\overline{M}$.

4143

In this demonstration we have not taken into consideration the emission of photons by propagators: indeed, it has been shown [3,4], that these terms do not lead to infra-red divergences.

4. – Separability of $\bar{M}$.

From eq. (9) we easily obtain for the matrix element of process (2), to the order e^{2m+s}

$$\bar{M}^{(2m+s)} = \frac{1}{m!}\left[\frac{\beta}{2}\int_{\lambda}^{\Delta\omega}\frac{dk}{k}\right]^m \langle f|\frac{(i)^s}{s!}T(L^s)|i\rangle_\lambda , \tag{13}$$

where

$$\beta = \frac{1}{2}\int d\Omega_k |k|^2 j_\mu(k) j_\mu^*(k)$$

and the auxiliary quantity λ, which has been introduced consistently in the two factors of eq. (13) as the lower limit to the permitted energies of real and virtual photons, has to become equal to zero.

Summing over m in the eq. (13), we have

$$\bar{M}^{(s)} = \sum_m \bar{M}^{(2m+s)} = \sum_m \frac{1}{m!}\left(\frac{\beta}{2}\right)^m\left(\log\frac{\Delta\omega}{\lambda}\right)^m \langle f|\frac{(i)^s}{s!}T(L^s)|i\rangle_\lambda$$

and then

$$\bar{M}^{(s)} = \left(\frac{\Delta\omega}{\lambda}\right)^{\beta/2}\langle f|\frac{(i)^s}{s!}T(L^s)|i\rangle_\lambda . \tag{14}$$

As eq. (14) holds for any s, by summing over s we finally have

$$\bar{M} = \langle f|\exp[-iA]S|i\rangle = \left(\frac{\Delta\omega}{\lambda}\right)^{\beta/2}\langle f|S|i\rangle_\lambda . \tag{15}$$

But, as we have shown in the preceeding Section that $\bar{M}$ is finite, the apparent divergence in λ of the eq. (15) when $\lambda\to 0$, will be cancelled by $\langle f|S|i\rangle_\lambda$ which consequently must go to zero when $\lambda\to 0$.

We can therefore write

$$\bar{M} = \left(\frac{\Delta\omega}{E}\right)^{\beta/2} M_E . \tag{16}$$

(3) D. R. Yennie and H. Suura: *Phys. Rev.*, **105**, 1378 (1957).
(4) D. R. Yennie, S. C. Frautschi and H. Suura: *Ann. of Phys.*, **13**, 379 (1961).

4144

The eq. (16) in which the auxiliary quantity λ does not appear any longer, defines formally M_E as a finite matrix element (whithout any infra-red divergence) for the perturbative series of process (1). Of course M_E may have ultraviolet divergences that, after renormalization, will give correction factors to the lowest-order matrix element for process (1).

The cross-section for process (2), will be proportional to $|\overline{M}|^2$

$$\text{d}\sigma \propto |\overline{M}|^2 = \left(\frac{\Delta\omega}{E}\right)^\beta |M_E|^2 = \left(\frac{\Delta\omega}{E}\right)^\beta \text{d}\sigma_E \,. \tag{17}$$

Actually the proportionality factor in eq. (17) can be completely determined if we take exactly account of the relation

$$\sum_{k=1}^{\infty} \omega_k \leqslant \Delta\omega \tag{18}$$

in defining final states.

5. – Correct expression of final states in an experiment with energy resolution $\Delta\omega$.

In order to determine the normalization resulting from (18) we shall adopt the following definition for the final states:

$$|f''\rangle = \frac{1}{2\pi}\int_{-\Delta\omega}^{\Delta\omega} \text{d}x \int_{-\infty}^{\infty} \exp[i\tau(H_s - x)]\,\text{d}\tau :\exp[iA_e]:|f\rangle \,, \tag{19}$$

where H_s represents the Hamiltonian of the electromagnetic field and where $j_\mu(k)$ is now defined by eq. (12).

The vectors $|f''\rangle$ in (19) represent the correct final states: in fact besides the detected particles, they contain also an indefinite number of photons with a total energy not greater than $\Delta\omega$.

The states (19) are not normalized: in order to achieve that, we calculate their square modulus. We have

$$\begin{aligned} N = {} & \frac{1}{4\pi^2}\int_{-\Delta\omega}^{\Delta\omega} \text{d}x \int_{-\Delta\omega}^{\Delta\omega} \text{d}x' \int_{-\infty}^{+\infty} \text{d}\tau \int_{-\infty}^{+\infty} \text{d}\tau' \langle f| :\exp[-iA_e]: \\ & :\exp[-i\tau(H_s - x)]\exp[i\tau'(H_s - x')] :\exp[iA]:|f\rangle_e = \\ = {} & \frac{1}{4\pi^2}\int_{-\Delta\omega}^{\Delta\omega} \text{d}x \int_{-\Delta\omega}^{\Delta\omega} \text{d}x' \int_{-\infty}^{+\infty} \text{d}\tau \int_{-\infty}^{+\infty} \text{d}\tau' \exp[-i(\tau x + \tau' x')]\cdot \\ & \cdot\exp\left[\int \text{d}^4k\,\delta(k^2)\theta(k_0)\,j_\mu(k)\,j^*_\mu(k)\times\exp[i(\tau - \tau')k_0]\right] = \frac{1}{\gamma^\beta}\left(\frac{\Delta\omega}{\lambda}\right)^\beta \frac{1}{\Gamma(1+\beta)}\,, \end{aligned} \tag{20}$$

4145

where $\gamma = e^C = 1.781$ is the Euler constant, β and λ were already defined above.

Now with the normalized states

$$(21)\qquad |f''\rangle = \frac{1}{\sqrt{N}}\frac{1}{2\pi}\int\limits_{-\Delta\omega}^{\Delta\omega}\mathrm{d}x\int\limits_{-\infty}^{+\infty}\mathrm{d}\tau\cdot\exp[i\tau(H_s - x)]:\exp[iA_0]:|f\rangle$$

it is possible to define a new matrix element M:

$$(22)\qquad M = \frac{1}{\sqrt{N}}\frac{1}{2\pi}\int\limits_{-\Delta\omega}^{\Delta\omega}\mathrm{d}x\int\limits_{-\infty}^{+\infty}\mathrm{d}\tau\langle f|:\exp[-iA_0]:\exp[-i\tau(H_s - x)]S|i\rangle$$

and M is such that $|M|^2$ gives directly the cross-section for process (2).

By proceeding as in Sect. **3** and **4**, we find for M the expression

$$(23)\qquad M = \frac{1}{\sqrt{N}}\left(\frac{\Delta\omega}{\lambda}\right)^{\beta/2}\frac{1}{\gamma_\beta}\frac{1}{\Gamma(1+\beta)}\langle f|S|i\rangle_\lambda$$

and taking into account eq. (20) we have

$$(24)\qquad M = \frac{1}{\gamma^{\beta/2}}\left(\frac{\Delta\omega}{\lambda}\right)^{\beta/2}\frac{1}{\sqrt{\Gamma(1+\beta)}}\langle f|S|i\rangle_\lambda\,.$$

By the same reasoning which led to eq. (16), we can write

$$(25)\qquad M = \frac{1}{\gamma^{\beta/2}}\frac{1}{\sqrt{\Gamma(1+\beta)}}\left(\frac{\Delta\omega}{E}\right)^{\beta/2}M_E$$

so that we finally obtain for the cross-section

$$(26)\qquad \mathrm{d}\sigma = \frac{1}{\gamma^\beta}\frac{1}{\Gamma(1+\beta)}\left(\frac{\Delta\omega}{E}\right)^{\beta}\mathrm{d}\sigma_E$$

where M_E and $\mathrm{d}\sigma_E$ are defined by eq. (17).

Equation (26) is in agreement with the results obtained by ETIM, PANCHERI and TOUSCHEK (5).

As we can see comparing eqs. (25) and (16), the only difference between them is the factor

$$C = \frac{1}{\gamma^{\beta/2}}\frac{1}{\sqrt{\Gamma(1+\beta)}}$$

(5) E. ETIM, G. PANCHERI and B. TOUSCHEK: Frascati, Internal Report LNF-66/38 (1966).

which for small values of β is $1-\pi^2\beta^2/24$, so it is very close to 1, as said before.

It can easily be shown that $(1/C^2)-1$ represents the probability that many photons, each with energy $<\Delta\omega$ combine to give a total energy loss $>\Delta\omega$.

Equation (25) obtained for the case of an experiment in which the momentum resolution is zero and only the energy resolution, $\Delta\omega\neq 0$ can be generalized if we replace the condition (18) by a more general condition which limits in some way also photon momenta: consequently eq. (15) too must be generalized. A method for doing his has been discussed in ref. ([5]).

* * *

It is a pleasure to thank Prof. B. TOUSCHEK for having suggested this work as well as for this constant assistance.

RIASSUNTO

In questo lavoro si mostra come la divergenza infrarossa può essere eliminata nell'elemento di matrice purché si usino, come stati finali, stati fisici veri per un esperimento in cui siano create e distrutte particelle cariche.

Замечание об инфракрасной расходимости.

Резюме (*). — В этой статье мы показываем, как возможно устранить инфракрасную расходимость из матричного элемента, при условии, что для эксперимента, включающего уничтожение и рождение заряженных частиц, используются физически правильные конечные состояния.

(*) *Переведено редакцией.*

Nuclear Physics B101 (1975) 234–262

RADIATIVE CORRECTIONS FOR COLLIDING BEAM RESONANCES

M. GRECO, G. PANCHERI-SRIVASTAVA * and Y. SRIVASTAVA * **
Laboratori Nazionali del CNEN, Frascati, Italy

Received 3 July 1975

Detailed expressions are presented for radiative corrections to colliding beam experiments in the presence of resonances, including interference effects. The derivation of our formulae is accomplished using perturbation theory methods as well as the coherent state formalism. These are then applied to determine the resonance parameters for $\psi(3.1)$ and $\psi(3.7)$.

1. Introduction

The problem of radiative corrections is well known and understood in quantum electrodynamics [1–3]. The physical origin lies in the fact that any scattering process involving charged particles in the initial or/and final state cannot take place without the emission of soft photons. The infrared divergence is artificially introduced in the perturbative solution of electrodynamics because of the unphysical separation of real from virtual soft photons, together with the unrealistic counting of the emitted quanta. The correct analysis of the collective soft-photon effect leads to the definition of an observable cross section $(\mathrm{d}\sigma)_{\mathrm{obs}}$, for an experiment with energy resolution $\Delta\omega$, which includes the infrared corrections in exponentiated form and is simply related to the lowest order cross section $\mathrm{d}\sigma_0$ by [1,3]

$$(\mathrm{d}\sigma)_{\mathrm{obs}} \propto \left(\frac{\Delta\omega}{E}\right)^{\beta} \mathrm{d}\sigma_0 \,, \tag{1.1}$$

where E is the energy of the radiating particle and $\beta \propto (4\alpha/\pi)\ln(2E/m)$ will be better defined later. Eq. (1.1) is a well known result which holds for any process whose bare cross section $\mathrm{d}\sigma_0$ is not a rapidly varying function of the energy or momentum transfer. It explicitly shows that the radiative correction depends upon an external parameter $\Delta\omega$, which characterizes a given experiment.

* Permanent Address: Northeastern University, Boston, Mass. 02115, USA.
** Supported in part by the National Science Foundation, USA.

In a scattering process which proceeds *via* the formation of a resonance, it is not possible to directly apply eq. (1.1), the reason being simply that the resonance cross section is very sensitive to an energy loss in the initial state, while no such effect results from photon emission from the final state. A first analysis of this problem has been performed in ref. [5], where eq. (1.1) has been generalized to

$$(\mathrm{d}\sigma)_{\mathrm{obs}} \propto \int_0^{\Delta\omega} \frac{\mathrm{d}\omega}{\omega}\left(\frac{\omega}{E}\right)^{\beta} \{\beta_{\mathrm{f}}\, \mathrm{d}\sigma_{\mathrm{R}}(2E) + \beta_{\mathrm{i}}\, \mathrm{d}\sigma_{\mathrm{R}}(2E-\omega)\}\,, \tag{1.2}$$

in the case of resonance formation in the *s*-channel ($\sqrt{s} = 2E$). Here, β_{f} and β_{i} are the radiative parameters of the final and initial state respectively, $\beta \simeq \beta_{\mathrm{f}} + \beta_{\mathrm{i}}$ and $\mathrm{d}\sigma_{\mathrm{R}}(\sqrt{s})$ refers to the typical (Breit–Wigner) resonant cross section. Eq. (1.2), which automatically reduces to (1.1) in the case of a smooth cross section, shows already a new and interesting feature, although providing only a partial answer to the problem (see below). In the case of a narrow resonance in fact ($\Gamma \ll \Delta\omega$), the width provides a natural cut-off in damping the integral over the energy loss in the initial state. Unlike eq. (1.1), one finds therefore that the soft photon emission is governed by an intrinsic physical quantity instead of the external parameter $\Delta\omega$, as far as emission from the initial state is concerned.

Interference effects have been neglected in eq. (1.2), i.e. the overlapping of radiation from the initial and final states. One would also like to have a full answer in the case of interference of a resonance term with a purely QED background. The recent discovery [6] of very narrow resonances decaying both into leptons and hadrons, provides the natural physical framework for these problems, demanding as well a precise evaluation of the radiative correction factors in order to extract from the experimental data the physical parameters of interest.

The aim of this work is to provide a detailed study of all these radiative effects in $\mathrm{e^+e^-}$ scattering and give a precise answer to those questions of immediate experimental interest. A brief account of our main results has already been given [7].

We consider the infrared factors to all orders in perturbation theory. This is achieved by using two different methods, both providing the same result. In the first case, as shown in sect. 3, we use a perturbative approach to all orders in the classical currents responsible for soft photon emission. This method is applied to the products of two matrix elements, in order to obtain the radiative correction factors to a purely resonant cross section as well as to the interference cross section between the resonance and a purely QED background. In sect. 4 we apply the coherent state formalism developed in ref. [8], directly to matrix elements which, unlike the usual perturbative expansion of QED, are finite and factorizable in the infrared factors. This directly follows from a more realistic definition of the final states, which contain an unlimited and undetermined number of soft photons.

The main issue in our results agrees with the qualitative arguments given after eq. (1.2) and can be simply stated by assigning to the initial state a "proper" energy loss $\Delta\omega_{\mathrm{i}}$, with a phase, given by

$$\Delta\omega_i = \Gamma/2 \sin\delta_R \, e^{i\delta_R} , \tag{1.3}$$

where the resonant amplitude is parametrized as $M_R \propto \sin\delta_R \, e^{i\delta_R}$. The energy loss from the final state $\Delta\omega_f$, on the contrary, coincides with the energy resolution $\Delta\omega$ of a given experiment. The overall radiative factors are then simply given in terms of suitable powers of $\Delta\omega_i$ and $\Delta\omega_f$, depending on the type of cross section involved.

Most questions of immediate experimental interest are studied in sect. 5, where we apply our formalism to actual experiments. In particular we discuss the case of experiments of resonance production with a machine resolution σ, such that $\sigma \gtrsim \Gamma$. The partial widths of the recently discovered $\psi(3.1)$ and $\psi(3.7)$ are then extracted from the experimental data. In this connection we give a simple formula which allows a careful estimate of the resonance parameters from a knowledge only of the peak values of the cross sections.

2. Notation and formulae

In this section, we define our notation and collect some relevant formulae.

For a process of the type

$$e^-(p_1) + e^+(p_2) \to A^-(p_3) + A^+(p_4)$$

we have:

$$s = W^2 = 4E^2 = -(p_1 + p_2)^2 , \tag{2.1a}$$

$$y = W - M , \tag{2.1b}$$

$$Z = \cos\vartheta = \hat{p}_1 \cdot \hat{p}_3 = \hat{p}_2 \cdot \hat{p}_4 , \tag{2.1c}$$

$$\beta_{i,f} = \frac{4\alpha}{\pi} \left[\ln \frac{W}{m_{i,f}} - \frac{1}{2} \right] , \tag{2.2a}$$

$$\beta_{int} = \frac{4\alpha}{\pi} \ln(\tan \tfrac{1}{2}\vartheta) , \tag{2.2b}$$

where $\Delta\omega$ is the energy resolution of the experiment, λ is the infrared cut off, m is the electron mass.

A resonance of mass M, and total width Γ may conveniently be parametrized in terms of a resonance phase-shift $\delta_R(W)$, defined as

$$\tan\delta_R(W) = \tfrac{1}{2}\Gamma/(-y) . \tag{2.3}$$

Cross sections for the processes $e^+e^- \to e^+e^-$, $\mu^+\mu^-$ and hadrons can be written as follows:

$$\sigma^c = C^{res}_{infra}\sigma_{res}(1 + C^{res}_F) + C^{int}_{infra}\sigma_{int}(1 + C^{int}_F) + C^{QED}_{infra}\sigma_{QED}(1 + C^{QED}_F) , \tag{2.4}$$

$$\left(\frac{d\sigma}{dZ}\right)^{\text{leptonic}}_{\text{res}} = \frac{\pi\alpha^2}{2s}(1+Z^2)\frac{\frac{1}{4}g^4M^2}{y^2+(\frac{1}{2}\Gamma)^2}, \tag{2.5a}$$

$$\left(\frac{d\sigma}{dZ}\right)^{\mu^+\mu^-}_{\text{int}} = \frac{\pi\alpha^2}{2s}(1+Z^2)\frac{g^2My}{y^2+(\frac{1}{2}\Gamma)^2}, \tag{2.5b}$$

$$\left(\frac{d\sigma}{dZ}\right)^{e^+e^-}_{\text{int}} = \frac{\pi\alpha^2}{2s}\frac{Z(3+Z^2)}{(1-Z)}\frac{-g^2My}{y^2+(\frac{1}{2}\Gamma)^2}, \tag{2.5c}$$

$$\left(\frac{d\sigma}{dZ}\right)^{\mu^+\mu^-}_{\text{QED}} = \frac{\pi\alpha^2}{2s}(1+Z^2), \tag{2.5d}$$

$$\left(\frac{d\sigma}{dZ}\right)^{e^+e^-}_{\text{QED}} = \frac{\pi\alpha^2}{2s}\left(\frac{3+Z^2}{1-Z}\right)^2, \tag{2.5e}$$

$$\sigma^{\text{hadron}}_{\text{res}} = \frac{\pi\alpha}{s}\frac{g^2M\Gamma_h}{y^2+(\frac{1}{2}\Gamma)^2}, \tag{2.5f}$$

where the coupling constant g is defined in terms of the leptonic widths (assumed equal) as

$$\Gamma_e \simeq \Gamma_\mu = \tfrac{1}{3}\alpha g^2 M, \qquad (\alpha^{-1} \approx 137), \tag{2.6}$$

and Γ_h is the hadronic width of the resonance.

The $C^{\text{res, int, QED}}_{\text{infra}}$ are the infrared factors associated with each of the respective cross sections, while $C^{\text{res,int,QED}}_{\text{F}}$ incorporate the rest of the finite radiative correction and must be calculated perturbatively. For a very narrow resonance, $\Gamma \ll \Delta\omega$, as shown in the text, the C_{infra} factors take the form

$$C^{\text{res}}_{\text{infra}} = \left(\frac{\Delta\omega}{E}\right)^{\beta_f}\left(\frac{\Gamma}{M}\right)^{\beta_i}\frac{\sin[\delta_R(1-\beta_i)]}{[\sin\delta_R]^{1+\beta_i}}\frac{1}{\gamma^{\beta_f}\Gamma(1+\beta_f)} \tag{2.7a}$$

$$\approx \left(\frac{\Delta\omega}{E}\right)^{\beta_f}\left(\frac{y^2+(\frac{1}{2}\Gamma)^2}{(\frac{1}{2}M)^2}\right)^{\beta_i/2}\left[1+\beta_i\frac{y}{\frac{1}{2}\Gamma}\delta_R\right], \tag{2.7b}$$

$$C^{\text{int}}_{\text{infra}} = \left(\frac{\Delta\omega}{E}\right)^{\beta_f+\beta_{\text{int}}}\left(\frac{\Gamma}{M}\right)^{\beta_i}\frac{\cos[\delta_R(1-\beta_i)]}{(\sin\delta_R)^{\beta_i}\cos\delta_R}\frac{1}{\gamma^{\beta_f+\beta_{\text{int}}}\Gamma(1+\beta_f+\beta_{\text{int}})} \tag{2.8a}$$

$$\approx \left(\frac{\Delta\omega}{E}\right)^{\beta_f+\beta_{\text{int}}}\left[\frac{y^2+(\frac{1}{2}\Gamma)^2}{(\frac{1}{2}M)^2}\right]^{\beta_i/2}[1-\beta_i(\tfrac{1}{2}\Gamma/y)\delta_R], \tag{2.8b}$$

$$C_{\text{infra}}^{\text{QED}} = \left(\frac{\Delta\omega}{E}\right)^{\beta_i+\beta_f+2\beta_{\text{int}}} \frac{1}{\gamma^{\beta_i+\beta_f+2\beta_{\text{int}}}\Gamma(1+\beta_i+\beta_f+2\beta_{\text{int}})}\,, \tag{2.9}$$

where $\ln\gamma = 0.5772$ is Euler's constant.

The β_{int} dependence in (2.7) and (2.8) is different from that shown in our earlier work [7]. This difference arises due to extra terms from the virtual graphs. (This is further discussed in the text and in appendix A).

The C_F factors are as follows:

2.1. $e^+e^- \to$ hadrons

Here one may safely neglect σ_{QED} and σ_{int} (at least for $\psi(3.1)$ and $\psi'(3.7)$ resonances). Then, we need only C_F^{res} for hadrons. It is [5,9]

$$C_F^{\text{hadron}} \approx \tfrac{13}{12}\beta_e + \frac{\alpha}{\pi}(\tfrac{1}{3}\pi^2 - \tfrac{17}{18})\,. \tag{2.10}$$

2.2. $e^+e^- \to \mu^+\mu^-$.

For this reaction, C_F^{QED} has been calculated in refs. [10–12]. The corresponding C_F^{int} and C_F^{res} are taken from ref. [13],

$$C_F^{\text{QED}} = \tfrac{13}{12}(\beta_e + \beta_\mu) + \frac{2\alpha}{\pi}(\tfrac{1}{3}\pi^2 - \tfrac{17}{18}) + X\,, \tag{2.11}$$

$$C_F^{\text{int}} = \tfrac{11}{12}(\beta_e + \beta_\mu) + \frac{2\alpha}{\pi}(\tfrac{1}{3}\pi^2 - \tfrac{13}{18}) + \tfrac{1}{2}X\,, \tag{2.12}$$

where

$$X = -\frac{4\alpha}{\pi}\left\{\frac{1}{1+Z^2}\left[Z\{(\ln\sin\tfrac{1}{2}\theta)^2 + (\ln\cos\tfrac{1}{2}\theta)^2\}\right.\right.$$
$$\left.+ (\sin^2\tfrac{1}{2}\theta)(\ln\cos\tfrac{1}{2}\theta) - (\cos^2\tfrac{1}{2}\theta)(\ln\sin\tfrac{1}{2}\theta)\right]$$
$$\left.+ \left[(\ln\cos\tfrac{1}{2}\theta)^2 - (\ln\sin\tfrac{1}{2}\theta)^2 + \tfrac{1}{2}\mathrm{Li}_2(\sin^2\tfrac{1}{2}\theta) - \tfrac{1}{2}\mathrm{Li}_2(\cos^2\tfrac{1}{2}\theta)\right]\right\}\,, \tag{2.13}$$

$$C_F^{\text{res}} = \tfrac{3}{4}(\beta_e + \beta_\mu) + \frac{2\alpha}{\pi}(\tfrac{1}{3}\pi^2 - \tfrac{1}{2}) + \tfrac{4}{9}\alpha\frac{\alpha\Gamma}{\Gamma_e} + Y\,, \tag{2.14}$$

where

$$Y = -\tfrac{2}{3}\alpha\frac{\alpha\Gamma}{\Gamma_e}\frac{1}{1+Z^2}\left[Z - 2Z(\ln\sin\tfrac{1}{2}\theta + \ln\cos\tfrac{1}{2}\theta) + 2(1+Z^2)\ln(\tan\tfrac{1}{2}\theta)\right]. \tag{2.15}$$

For all practical purpose the approximation $C_F^{\text{int}} \approx \tfrac{1}{2}(C_F^{\text{QED}} + C_F^{\text{res}})$ works quite well, since Γ-dependent terms are quite small.

2.3. $e^+e^- \to e^+e^-$

The C_F^{res} for this reaction may be read off directly from eq. (2.14) substituting $2\beta_e$ for the factor $(\beta_e + \beta_\mu)$ appearing there. The C_F^{QED} is obtained from ref. [12],

$$C_F^{QED} = \frac{\alpha}{\pi}\,[u^2 - v^2 + w^2 + 2\mathrm{Li}_2(\tfrac{1}{2}(1+Z)) - 2\mathrm{Li}_2(\tfrac{1}{2}(1-Z)) - \tfrac{2}{3}\pi^2]$$

$$+\frac{\alpha}{\pi}\,\frac{1}{(3+Z^2)^2}\,[\tfrac{1}{3}u(3-42Z+42Z^2-14Z^3+11Z^4) - v(5-7Z+3Z^2-Z^3)$$

$$+\tfrac{1}{3}w(111+21Z+33Z^2+11Z^3) + \tfrac{1}{2}u^2(3+7Z-5Z^2-3Z^3-2Z^4)$$

$$+v^2(3-3Z+Z^2-Z^3) - \tfrac{1}{2}w^2(9+7Z+11Z^2+5Z^3) - 2uvZ(2-Z-Z^3)$$

$$-uw(21+3Z+9Z^2-3Z^3+2Z^4) + 2vw(6+5Z+4Z^2+Z^3)$$

$$-\tfrac{46}{9}(9+6Z^2+Z^4) + \tfrac{1}{3}\pi^2(18-15Z+12Z^2-3Z^3+4Z^4)]\,, \tag{2.16}$$

where

$$u = 2\ln\frac{W}{m}\,, \qquad v = \ln\frac{2E^2(1+Z)}{m^2}\,, \qquad w = \ln\frac{2E^2(1-Z)}{m^2}\,, \tag{2.17}$$

and the dilogarithm $\mathrm{Li}_2(x)$ is defined as

$$\mathrm{Li}_2(x) = -\int_0^x \frac{\ln(1-y)}{y}\,\mathrm{d}y\,. \tag{2.18}$$

The C_F^{int} expression is missing. However, as discussed before, the approximation $C_F^{int} \approx \tfrac{1}{2}(C_F^{QED} + C_F^{res})$ ought to work well.

This completes the list of our formulae. Their derivation, discussion and application follows in the coming sections.

3. Perturbative method

Let us consider the process

$$e^-(p_1) + e^+(p_2) \to A^-(p_3) + A^+(p_4)\,, \tag{3.1}$$

which we first assume to be described by a smooth function of the energy. In this case process (3.1), accompanied by the emission of n real soft photons can be described by [2,3]

$$M^{(n)}_{\mu_1\ldots\mu_n} \simeq j_{\mu_1}(k_1)\ldots j_{\mu_n}(k_n)\,M(p_1,p_2,p_3,p_4)\,, \tag{3.2}$$

where $M(p_1, p_2, p_3, p_4)$ is the matrix element for process (1) including all the virtual processes, and

$$j_\mu(k) = \frac{ie}{(2\pi)^{3/2}}\left[\frac{p_{1\mu}}{p_1\cdot k} - \frac{p_{2\mu}}{p_2\cdot k} - \frac{p_{3\mu}}{p_3\cdot k} + \frac{p_{4\mu}}{p_4\cdot k}\right].$$

If process (3.1) proceeds via a narrow resonance, the matrix element, which describes it, is a rapidly varying function of the energy. Eq. (3.2) then is not valid, since soft photon emission from the initial state shifts the position of the resonance and prevents a complete separation of the infrared terms from the unperturbed matrix element. It is still possible however to have a working analogue of eq. (3.2) in the following way. Let us first consider the emission of just one real soft photon in process (3.1). We can write

$$M_\mu^{(1)} \simeq I_\mu(k)\, M(W-k) + F_\mu(k) M(W)\,, \tag{3.3}$$

where we have distinguished between initial and final state emission on the basis that one shifts the c.m. energy W to a lower value $W-k$, while the other does not. In eq. (3.3) $M(W)$ or $M(W-k)$ is the resonant matrix element inclusive of all virtual processes and

$$I_\mu(k) = \frac{ie}{(2\pi)^{3/2}}\left[\frac{p_{1\mu}}{p_1\cdot k} - \frac{p_{2\mu}}{p_2\cdot k}\right],$$

$$F_\mu(k) = \frac{ie}{(2\pi)^{3/2}}\left[\frac{-p_{3\mu}}{p_3\cdot k} + \frac{p_{4\mu}}{p_4\cdot k}\right].$$

We now parametrize the amplitudes $M(W)$ and $M(W-k)$ by writing

$$M(W-k) = \int \mathrm{d}\omega\, \delta(\omega - k)\, M(W-\omega)\,,$$

$$M(W) = \int \mathrm{d}\omega\, \delta(\omega)\, M(W-\omega)\,,$$

and use the representation

$$\delta(x) = \frac{1}{2\pi}\int_{-\infty}^{+\infty} \mathrm{d}t\, \mathrm{e}^{ixt}\,,$$

so that eq. (3.3) becomes

$$M_\mu^{(1)} \simeq \int \mathrm{d}\omega\, M(W-\omega) \int_{-\infty}^{+\infty} \frac{\mathrm{d}t}{2\pi}\, \mathrm{e}^{i\omega t}\, \tilde{j}_\mu(k,t)\,, \tag{3.4}$$

with

$$\tilde{j}_\mu(k,t) = I_\mu(k)\, \mathrm{e}^{-ikt} + F_\mu(k)\,.$$

In eq. (3.4) the limits of integration in ω will be fixed later by the requirement that the total energy loss be less than or equal to the energy resolution $\Delta\omega$ of the experiment.

We can now iterate eq. (3.4) to obtain the matrix element corresponding to the emission of n real soft photons

$$M^{(n)}_{\mu_1 \cdots \mu_n} = \int \mathrm{d}\omega\, M(W-\omega) \int_{-\infty}^{+\infty} \frac{\mathrm{d}t}{2\pi}\, \mathrm{e}^{i\omega t}\, \tilde{j}_{\mu_1}(k_1, t) \ldots \tilde{j}_{\mu_n}(k_n, t)\,. \tag{3.5}$$

The cross section for the nth order bremsstrahlung is then proportional to

$$\frac{1}{n!}\int \mathrm{d}\omega\, M(W-\omega) \int \mathrm{d}\omega'\, M^{+}(W-\omega') \int_{-\infty}^{+\infty} \frac{\mathrm{d}t}{2\pi}\, \mathrm{e}^{i\omega t} \int_{-\infty}^{+\infty} \mathrm{e}^{-i\omega' t'} \frac{\mathrm{d}t'}{2\pi}$$
$$\times \left[\int_0^{\Delta\omega} \frac{\mathrm{d}^3 k}{2k_0}\, \tilde{j}_\mu(k, t)\, \tilde{j}^{\mu *}(k, t')\right]^n .$$

Summing over n we get the cross section for process (3.1) accompanied by a radiation loss up to $\Delta\omega$

$$\sigma_{\mathrm{res}}(W, \theta) \propto \int \mathrm{d}\omega\, M(W-\omega) \int \mathrm{d}\omega'\, M^{+}(W-\omega') \int_{-\infty}^{+\infty} \frac{\mathrm{d}t}{\mathrm{d}\pi}\, \mathrm{e}^{i\omega t}$$
$$\times \int_{-\infty}^{+\infty} \frac{\mathrm{d}t'}{2\pi}\, \mathrm{e}^{-i\omega' t'}\, \mathrm{e}^{h(t,t')}\,, \tag{3.6}$$

where

$$h(t, t') = \int \frac{\mathrm{d}^3 k}{2k_0}\, \tilde{j}_\mu(k, t)\, \tilde{j}^{\mu *}(k, t')$$
$$= \beta_{\mathrm{i}} \int_\lambda^{\Delta\omega} \frac{\mathrm{d}k}{k}\, \mathrm{e}^{-ik(t-t')} + \beta_{\mathrm{int}} \int_\lambda^{\Delta\omega} \frac{\mathrm{d}k}{k}\, (\mathrm{e}^{-ikt} + \mathrm{e}^{ikt'}) + \beta_{\mathrm{f}} \int_\lambda^{\Delta\omega} \frac{\mathrm{d}k}{k}\,, \tag{3.7}$$

where

λ = infrared energy cut-off,

$\Delta\omega$ = energy resolution of the experiment,

$$\beta_{\mathrm{i}} = \tfrac{1}{2}k^2 \int \mathrm{d}\Omega_k\, (I_\mu I^{\mu *}) \simeq \frac{4\alpha}{\pi}\left(\ln \frac{W}{m_{\mathrm{i}}} - \frac{1}{2}\right),$$

$$\beta_{\mathrm{int}} = \tfrac{1}{2}k^2 \int \mathrm{d}\Omega_k\, (I_\mu F^{\mu *}) \simeq \frac{4\alpha}{\pi} \ln (\mathrm{tg}\, \tfrac{1}{2}\theta)\,,$$

$$\beta_{\mathrm{f}} = \tfrac{1}{2}k^2 \int \mathrm{d}\Omega_k\, (F_\mu F^{\mu *}) \simeq \frac{4\alpha}{\pi}\left(\ln \frac{W}{m_{\mathrm{f}}} - \frac{1}{2}\right),$$

θ = c.m. scattering angle between equally charged particles.

The photon energy being positive definite, the ω and ω' integrals in eq. (3.6) extend only over positive values (this is also a consequence of the analyticity properties of $h(t, t')$).

In eq. (3.7) we have introduced a photon minimum energy to handle the infrared divergence arising from the description of real photon emission. We can rewrite eq. (3.7) as

$$h(t, t') = (\beta_\mathrm{i} + 2\beta_\mathrm{int} + \beta_\mathrm{f}) \ln \frac{\Delta\omega}{\lambda} + f(t, t') ,$$

where now $f(t, t')$ is finite as $\lambda \to 0$ and

$$f(t, t') = \beta_\mathrm{i} \int_0^{\Delta\omega} \frac{\mathrm{d}k}{k} (\mathrm{e}^{-ik(t-t')} - 1) + \beta_\mathrm{int} \int_0^{\Delta\omega} \frac{\mathrm{d}k}{k} (\mathrm{e}^{-ikt} + \mathrm{e}^{ikt'} - 2) . \qquad (3.8)$$

Eq. (3.6) thus becomes

$$\sigma_\mathrm{res}(W, \theta) \propto \left(\frac{\Delta\omega}{\lambda}\right)^\beta \int \mathrm{d}\omega\, M(W-\omega) \int \mathrm{d}\omega'\, M^+(W-\omega') \int_{-\infty}^{+\infty} \frac{\mathrm{d}t}{2\pi}\, \mathrm{e}^{i\omega t}$$

$$\times \int_{-\infty}^{+\infty} \frac{\mathrm{d}t'}{2\pi}\, \mathrm{e}^{-i\omega' t'}\, \mathrm{e}^{f(t,t')} , \qquad (3.9)$$

where $\beta = \beta_\mathrm{i} + 2\beta_\mathrm{int} + \beta_\mathrm{f}$. If we assume $\beta_\mathrm{int} = 0$, the t and t' integrations can easily be done. We postpone to the next section and to appendix A the discussion on the β_int dependence of σ_res and proceed by putting $\beta_\mathrm{int} = 0$. Then $f(t, t')$ depends only upon the variable $\tau = t - t'$ so that eq. (3.9) can be rewritten as

$$\sigma_\mathrm{res}(W, \theta) \propto \left(\frac{\Delta\omega}{\lambda}\right)^{\beta_\mathrm{i}+\beta_\mathrm{f}} \int \mathrm{d}\omega\, M(W-\omega) \int \mathrm{d}\omega'\, M^+(W-\omega')$$

$$\times \int_{-\infty}^{+\infty} \frac{\mathrm{d}T}{2\pi}\, \mathrm{e}^{i(\omega-\omega')T} \int_{-\infty}^{+\infty} \frac{\mathrm{d}\tau}{2\pi}\, \mathrm{e}^{i(\omega+\omega')/2\tau}\, \mathrm{e}^{f(\tau)} .$$

The $\mathrm{d}T$ integral gives a $\delta(\omega - \omega')$ and we are left with

$$\sigma_\mathrm{res}(W, \theta) \propto \left(\frac{\Delta\omega}{\lambda}\right)^{\beta_\mathrm{i}+\beta_\mathrm{f}} \int \frac{\mathrm{d}\omega}{\omega} |M(W-\omega)|^2 \int_{-\infty}^{+\infty} \frac{\mathrm{d}\tau}{2\pi}\, \mathrm{e}^{i\tau\omega}$$

$$\times \exp\left[\beta_\mathrm{i} \int_0^{\Delta\omega/\omega} \frac{\mathrm{d}k}{k} (\mathrm{e}^{-ik\tau} - 1)\right] . \qquad (3.10)$$

The apparent λ divergence in eq. (3.10) must be cancelled by the virtual photon contributions, i.e. it must be

$$|M(W-\omega)|^2 = \left(\frac{\lambda}{E}\right)^{\beta_i+\beta_f} |M_E(W-\omega)|^2 , \tag{3.11}$$

with $E = \frac{1}{2}W$ and $|M_E(W-\omega)|^2$ containing, upon renormalization, only finite contributions. In $M_E(W-\omega)$, the subscript E indicates the scale used to extract the virtual photon contribution to the λ divergence. Inserting eq. (3.11) into eq. (3.10) and performing the $d\tau$ integration as in ref. [4] we finally obtain

$$\sigma_{res}(W,\theta) = \left(\frac{\Delta\omega}{E}\right)^{\beta_f} N(\beta_i) \frac{\sin(\pi\beta_i)}{\pi} \int_0^{\Delta\omega} \frac{d\omega}{\omega} \left(\frac{\omega}{E}\right)^{\beta_i} \sigma_E(W-\omega,\theta) , \tag{3.12}$$

where $N(\beta_i)$ is a normalization factor which is related to the probability that two or more photons from the initial state combine to give a total energy loss larger than $\Delta\omega$. Since $N(\beta_i) - 1$ is of order β_i^2 we can safely set it equal to 1. It should be noticed that in eq. (3.12) no similar normalization factor for final state emission appears. This reflects the fact that in obtaining this equation we have not applied energy conservation to the emission from the final state. The error involved is obviously of order β_f^2 and hence negligible.

In eq. (3.12) the maximum energy loss from the initial state has been set equal to that from the final state. In an actual e^+e^- colliding beam experiment this, in general, is not correct: in the ADONE experiment, for instance, the energy resolution of the initial state is of order 100 MeV (for $W \sim 3$ GeV) while that of the final state may be four or five times larger. However, for resonant processes characterized by a width $\Gamma \ll \Delta\omega$ and in the resonance region, the $d\omega$ integral is highly insensitive to the upper limit. To see this, we must first extract the dependence on $\sigma_E(W-\omega,\theta)$. From ref. [5] we have to the next order in α,

$$\sigma_E(W-\omega,\theta) \simeq \frac{A_0(\frac{1}{2}\Gamma)^2 \, [1 + C_F^{res}]}{(W-M-\omega)^2 + (\frac{1}{2}\Gamma)^2} , \tag{3.13}$$

where C_F^{res} (see sect. 2) is a finite contribution of order α, independent of ω. The term A_0 depends on the specific process under consideration and represents the peak value of the unshifted resonant cross section, i.e.

$$\sigma_0(W,\theta) = \frac{A_0(\frac{1}{2}\Gamma)^2}{(W-M)^2 + (\frac{1}{2}\Gamma)^2} .$$

We can now insert eq. (3.13) into (3.12) and evaluate the integral. As it is, the integration depends upon the resonance parameters and gives different results for $\Delta\omega \gg \Gamma$ or $\Delta\omega \ll \Gamma$. In the case of the $\psi(3.1)$ and $\psi(3.7)$ resonances, we can safely set the upper limit of integration equal to ∞. The error involved is of order

$$\beta_i \frac{(W-M)^2 + (\frac{1}{2}\Gamma)^2}{\Delta\omega^2} ,$$

hence negligible. We thus get, for a very narrow resonance,

$$\sigma_{\rm res}(W,\theta)=\left(\frac{\Delta\omega}{E}\right)^{\beta_{\rm f}}\left[\frac{(W-M)^2+(\frac{1}{2}\Gamma)^2}{(\frac{1}{2}M)^2}\right]^{\beta_{\rm i}/2}\frac{\sin\left[\delta_{\rm R}(1-\beta_{\rm i})\right]}{\sin\delta_{\rm R}}$$

$$\times\,\sigma_0(W,\theta)\,(1+C_{\rm F}^{\rm res})\,, \qquad (3.14)$$

where $\delta_{\rm R}$ is the resonant phase shift, i.e. $\mathrm{ctg}\,\delta_{\rm R}=(M-W)/\frac{1}{2}\Gamma$.

We now wish to comment briefly on the various factors appearing in the r.h. side of eq. (3.14):

(a) The $C_{\rm F}^{\rm res}$ represents finite virtual photon contributions as well as, eventually, hard bremsstrahlung contributions from the final state. To determine it, one should proceed as follows. A first order expansion of eq. (3.14) gives

$$\sigma_{\rm res}(W,\theta)\simeq\left[1+\beta_{\rm f}\ln\frac{\Delta\omega}{E}+\tfrac{1}{2}\beta_{\rm i}\ln\frac{(W-M)^2+(\frac{1}{2}\Gamma)^2}{(\frac{1}{2}M)^2}\right.$$

$$\left.+\beta_{\rm i}\frac{W-M}{\frac{1}{2}\Gamma}\,\delta_{\rm R}+C_{\rm F}^{\rm res}\right]\sigma_0(W,\theta)\,.$$

By comparing the above expansion with the QED perturbative calculation to the same order one can then determine $C_{\rm F}^{\rm res}$. This was done, for instance in ref. [5] for the process $e^+e^-\to\rho\to\pi^+\pi^-$. The $C_{\rm F}^{\rm res}$ for the processes of interest for ψ production are given in sect. 2.

(b) The factor

$$\left[\frac{(W-M)^2+(\frac{1}{2}\Gamma)^2}{(\frac{1}{2}M)^2}\right]^{\beta_{\rm i}/2}$$

shows that the emission of photons from the initial state is regulated by the width of the resonance. This agrees with our physical intuition: if an electron–positron pair of total c.m. energy $W\simeq M$ radiates an energy larger than Γ the subsequent annihilation will not be observed as a resonant process. The resonant peak cross section is therefore reduced by a factor $(\Gamma/M)^{\beta_{\rm i}}$.

(c) The factor

$$\frac{\sin\left[\delta_{\rm R}(1-\beta_{\rm i})\right]}{\sin\delta_{\rm R}}$$

gives the radiative tail of the resonance. In fact an expansion in $\beta_{\rm i}$ ($\beta_{\rm i}=0.076$ at $W=3$ GeV) gives

$$\frac{\sin\left[\delta_{\rm R}(1-\beta_{\rm i})\right]}{\sin\delta_{\rm R}}\simeq1+\beta_{\rm i}\frac{W-M}{\frac{1}{2}\Gamma}\,\delta_{\rm R}\,,$$

which shows the characteristic radiative tail on the right of the resonance where $\delta_R \geqslant \frac{1}{2}\pi$. This is mostly a single photon effect, unlike the previous one.

Eq. (3.14) gives the radiatively corrected resonant part of the cross section. When the QED background cannot be neglected, interference effects between the resonant amplitude and the photon channels should be taken into account. In the following we derive the expression for the radiative corrections to the interference part of the cross section. To obtain the correction factor we must consider the interference between the QED amplitude for process (3.1) accompanied by the emission of n real soft photons, with the analogous resonant amplitude and then sum over n. The QED amplitude is given by eq. (3.2), i.e.

$$M^{(\mathrm{QED})}_{\mu_1 \ldots \mu_n} \simeq j_{\mu_1}(k_1) \ldots j_{\mu_n}(k_n)\, M^{(\mathrm{QED})}(W) \,,$$

and the resonant one by eq. (3.5), i.e.

$$M^{(\mathrm{res})}_{\mu_1 \ldots \mu_n} \simeq \int \mathrm{d}\omega\, M^{(\mathrm{res})}(W-\omega) \int_{-\infty}^{+\infty} \frac{\mathrm{d}t}{2\pi}\, \mathrm{e}^{i\omega t}\, \tilde{j}_{\mu_1}(k_i, t) \ldots \tilde{j}_{\mu_n}(k_n, t) \,,$$

where $M^{(\mathrm{QED})}(W)$ and $M^{(\mathrm{res})}(W-\omega)$ are the elastic amplitudes for process (3.1) inclusive of all virtual processes. One gets for the radiatively corrected interference cross section

$$\sigma_{\mathrm{int}}(W, \theta) \propto 2\,\mathrm{Re}\ [M^{(\mathrm{QED})}(W)]^+ \int \mathrm{d}\omega\, M^{(\mathrm{res})}(W-\omega) \int_{-\infty}^{+\infty} \frac{\mathrm{d}t}{2\pi}\, \mathrm{e}^{\mathrm{i}\omega t}\, \mathrm{e}^{g(t)} \,, \tag{3.15}$$

with

$$g(t) = (\beta_\mathrm{i} + 2\beta_{\mathrm{int}} + \beta_\mathrm{f}) \ln \frac{\Delta\omega}{\lambda} + (\beta_\mathrm{i} + \beta_{\mathrm{int}}) \int_0^{\Delta\omega} \frac{\mathrm{d}k}{k} (\mathrm{e}^{-ikt} - 1) \,.$$

In $g(t)$ we have separated out the infrared divergence introducing, as before, a minimum photon energy λ. The analyticity properties of $g(t)$ restrict the ω integration to positive values. For $\omega < \Delta\omega$ eq. (3.15) gives

$$\sigma_{\mathrm{int}}(W, \theta) \propto 2\,\mathrm{Re} \left\{ [M^{\mathrm{QED}}(W)]^+ \left(\frac{\Delta\omega}{\lambda}\right)^{\beta} \left[\frac{\sin[\pi(\beta_\mathrm{i} + \beta_{\mathrm{int}})]}{\pi}\right] \right\}$$

$$\times \int_0^{\Delta\omega} \frac{\mathrm{d}\omega}{\omega} \left(\frac{\omega}{\Delta\omega}\right)^{\beta_\mathrm{i} + \beta_{\mathrm{int}}} M^{\mathrm{res}}(W-\omega) \,.$$

As before we put $\beta_{\mathrm{int}} = 0$ and cancel the λ divergence by postulating that (see appendix A)

$$2\,\mathrm{Re}\,\{[M^{\mathrm{QED}}(W)]^+\,M^{\mathrm{res}}(W-\omega)\}$$

$$=\left(\frac{\lambda}{E}\right)^{\beta_i+\beta_f} 2\,\mathrm{Re}\,\{[M_{\mathrm{E}}^{\mathrm{QED}}(W)]^+\,M_{\mathrm{E}}^{\mathrm{res}}(W-\omega)\}\ ,$$

so that

$$\sigma_{\mathrm{int}}(W,\theta)=\left(\frac{\Delta\omega}{E}\right)^{\beta_f}\frac{\sin(\pi\beta_i)}{\pi}\int_0^{\Delta\omega}\frac{\mathrm{d}\omega}{\omega}\left(\frac{\omega}{E}\right)^{\beta_i}\sigma_{\mathrm{E}}^{\mathrm{int}}(W-\omega,\theta)\ . \tag{3.16}$$

We now have to extract the residual ω dependence on $\sigma_{\mathrm{E}}^{\mathrm{int}}(W-\omega)$. We can write

$$\sigma_{\mathrm{E}}^{\mathrm{int}}(W-\omega,\theta)\simeq\frac{\frac{1}{2}B_0\Gamma(W-M-\omega)[1+C_{\mathrm{F}}^{\mathrm{int}}]}{(W-M-\omega)^2+(\frac{1}{2}\Gamma)^2}\ , \tag{3.17}$$

where $C_{\mathrm{F}}^{\mathrm{int}}$, as before, is a finite contribution of order α, independent of ω. The B_0 is defined so that the interference cross section between the Breit-Wigner and the lowest order QED amplitude is

$$\sigma_0^{\mathrm{int}}(W,\theta)=\frac{\frac{1}{2}B_0(W-M)\Gamma}{(W-M)^2+(\frac{1}{2}\Gamma)^2}\ .$$

The insertion of eq. (3.17) into (3.16) gives

$$\sigma_{\mathrm{int}}(W,\theta)=\left(\frac{\Delta\omega}{E}\right)^{\beta_f}\left[\frac{(W-M)^2+(\frac{1}{2}\Gamma)^2}{(\frac{1}{2}M)^2}\right]^{\beta_i/2}$$

$$\times\frac{\cos[\delta_{\mathrm{R}}(1-\beta_i)]}{\cos\delta_{\mathrm{R}}}\ \sigma_0^{\mathrm{int}}(W,\theta)\,(1+C_{\mathrm{F}}^{\mathrm{int}})\ , \tag{3.18}$$

having let $\Delta\omega\to\infty$, as before.

In eq. (3.18) the term

$$\frac{\cos[\delta_{\mathrm{R}}(1-\beta_i)]}{\cos\delta_{\mathrm{R}}}\simeq\frac{-\frac{1}{2}\beta_i\Gamma}{W-M}\ \delta_{\mathrm{R}}+1$$

shows that the interference is not zero at the peak, where $\delta_{\mathrm{R}}=\frac{1}{2}\pi$. This is due to the fact that photon emission from the initial state generates, at the peak, a real part in the resonant amplitude. It is mostly a single photon effect, just like the radiative tail in σ_{res}.

4. Coherent state approach

4.1. Pure QED processes

The application of the coherent state formalism to the problem of the infrared divergence in electrodynamics was developed in ref. [8]. For the reader's convenience, and with one eye to its generalization for resonant processes given below, we briefly review the results obtained in ref. [8].

As already noted above, the infrared divergence in QED is entirely due to the unphysical description of the perturbative solution. The simple observation that the number of soft photons emitted in a reaction can never be measured, immediately leads to the necessity of introducing a new definition of final states, closer to physical reality. Consider a process

$$a + b \to c + d + \ldots, \tag{4.1}$$

in which at least some of the particles a, b, c, d, ... involved are charged. What one really observes is the process

$$a + b \to c + d + \ldots + \Gamma, \tag{4.2}$$

where Γ stands for an unlimited number of photons subject only to the condition that the total 4-momentum of the emitted photons is fixed by the energy and momentum resolution of the experimental apparatus. In the following, for the sake of simplicity, we shall assume only an energy resolution $\Delta\omega$. By calling $|f\rangle$ the final state of the reaction (4.1), let us introduce the state vector $|f'\rangle$, defined as

$$|f'\rangle = e^{i\Lambda_c}|f\rangle, \tag{4.3}$$

with

$$\Lambda_c = \int d^4x\, j_\mu(x)\, A_\mu(x), \tag{4.4}$$

where $A_\mu(x)$ is the quantized electromagnetic field, and $j_\mu(x)$ the Fourier transform of the classical current $j_\mu(k)$ defined as

$$j_\mu(k) = \frac{ie}{(2\pi)^{3/2}} \sum_{i=a,b,\ldots} \epsilon_i \frac{p_\mu^{(i)}}{(p^i\cdot k)} \qquad \text{for } k_0 \leqslant \Delta\omega,$$

$$j_\mu(k) = 0 \qquad \text{for } k_0 > \Delta\omega, \tag{4.5}$$

with $\epsilon_i = +1$ for positive outgoing and negative incoming particles and -1 otherwise. The operator $e^{i\Lambda_c}$ defined in eq. (4.3) is unitary. The state $|f'\rangle$ contains an unlimited and undetermined number of photons, created by the electromagnetic field of the distribution of classical currents (4.5).

It is now possible to define a matrix element $\overline{M} = \langle f'|S|i\rangle$ which, as proved in ref. [8] is: (i) finite and does not therefore possess an infrared divergence, (ii) separable in the infrared factor and (iii) directly comparable with the experimental cross section which results proportional to $|\bar{M}|^2$. In particular one has

$$\bar{M} = \langle \mathrm{f} |\, \mathrm{e}^{-i\Lambda_{\mathrm{c}}} S | \mathrm{i} \rangle = \exp\left\{ \frac{1}{2} \int_{\lambda}^{\Delta\omega} \frac{\mathrm{d}^3 k}{2k} \,[j_\mu(k)\, j^*_\mu(k)] \right\} \langle \mathrm{f} | S | \mathrm{i} \rangle_\lambda$$

$$= \exp\left\{ \tfrac{1}{2}\beta \int_{\lambda}^{\Delta\omega} \frac{\mathrm{d}k}{k} \right\} \langle \mathrm{f} | S | \mathrm{i} \rangle_\lambda = \left(\frac{\Delta\omega}{\lambda}\right)^{\beta/2} \langle \mathrm{f} | S | \mathrm{i} \rangle_\lambda \,, \tag{4.6}$$

where

$$\beta = \tfrac{1}{2} \int \mathrm{d}\Omega_k \,|k|^2 \, j_\mu(k)\, j^*_\mu(k)\,.$$

The auxiliary parameter λ has been introduced consistently both in

$$\int_{\lambda} [j_\mu(k)\, j^*_\mu(k)]\; \mathrm{d}^3 k/2k$$

and $\langle \mathrm{f} | S | \mathrm{i} \rangle_\lambda$ as the lower limit of the permitted energies of real and virtual photons, and has to be finally set to zero. In this limit, because of the finiteness of $\bar{M}$, the apparent divergence in $(\Delta\omega/\lambda)^{\beta/2}$ is cancelled by an analogous divergence in $\langle \mathrm{f} | S | \mathrm{i} \rangle_\lambda$. One can therefore write

$$\bar{M} = \left(\frac{\Delta\omega}{E}\right)^{\beta/2} M_{\mathrm{E}} \,, \tag{4.7}$$

which finally defines a matrix element M_{E} which is finite, i.e. without any infrared divergence. Of course M_{E} may have ultraviolet divergence which, upon normalization, will give additive correction factors.

The cross section for process (4.2) is proportional to $|\bar{M}|^2$,

$$(\mathrm{d}\sigma)_{\mathrm{obs}} \propto |\bar{M}|^2 = \left(\frac{\Delta\omega}{E}\right)^{\beta} |M_{\mathrm{E}}|^2 \propto \left(\frac{\Delta\omega}{E}\right)^{\beta} \mathrm{d}\sigma_{\mathrm{E}} \,, \tag{4.8}$$

and comparison with perturbation theory finally determines $\mathrm{d}\sigma_{\mathrm{E}}$. In $\mathrm{e}^+\mathrm{e}^-$ scattering $(\mathrm{e}^+\mathrm{e}^- \to \mathrm{f})$, one has for example

$$(\mathrm{d}\sigma)_{\mathrm{obs}} = \frac{1}{\gamma^\beta} \frac{1}{\Gamma(1+\beta)} \left(\frac{\Delta\omega}{E}\right)^{\beta} \mathrm{d}\sigma_{\mathrm{E}} = \frac{1}{\gamma^\beta} \frac{1}{\Gamma(1+\beta)} \left(\frac{\Delta\omega}{E}\right)^{\beta} \mathrm{d}\sigma_0 (1 + C_{\mathrm{F}})\,, \tag{4.9}$$

where $\sqrt{s} = 2E$ is the total energy in the c.m. frame, C_{F} is the left-over finite correction and $\mathrm{d}\sigma_0$ is the lowest order cross section for the process of interest. The extra factor $\gamma^\beta \Gamma(1+\beta)$, where $\ln \gamma = 0.5772$ is Euler's constant, is found [8], as also discussed in appendix B, by imposing the global condition

$$\sum_{n=1}^{\infty} k_0^{(n)} \leqslant \Delta\omega \tag{4.10}$$

on the states $|f'\rangle$, instead of $k_0^{(n)} \leqslant \Delta\omega$ for every n, as implied in (4.5). This factor however, for small values of β is $1 - \pi^2\beta^2/12$, so it is very close to 1. This is related to the fact that $[\gamma^\beta\Gamma(1+\beta) - 1]$ represents the probability that two or more photons, each with energy $<\Delta\omega$ combine to give a total energy loss $>\Delta\omega$.

4.2. Pure resonant processes

Eqs. (4.7)–(4.9) are valid provided the "bare" matrix element (without soft photon emission) is not a rapidly varying function of the energy or the momentum transfer. Let us now consider the modifications to our formalism due to the presence of a resonance, for example in the reaction $e^+e^- \to$ resonance $\to$ f.

As shown in ref. [5] it can be easily seen that the presence of a resonance, of mass M and width Γ, only modifies the classical current relative to the initial state, as follows:

$$j_\mu^R(k) \doteq \frac{(W-M) + \frac{1}{2}i\Gamma}{(W-M-k) + \frac{1}{2}i\Gamma}\, I_\mu(k) + F_\mu(k)\,, \qquad (4.11)$$

where, in the notations of the preceding section, $I_\mu(k)$ and $F_\mu(k)$ are the classical currents relative to the initial and final states respectively, in absence of the resonance The current $j_\mu^R(k)$ is exactly the Fourier transform of

$$j^R(x) = j^f(x)\,\theta(t) + \overline{j^{(i)}}(x)\,, \qquad (4.12)$$

where $\overline{j^{(i)}}(x)$ is given by

$$\overline{j^{(i)}}(x) = (-iE^*) \int_0^\infty dt_0\; j^{(i)}(x)\,\theta(-t-t_0)\, e^{iE^* t_0}\,, \qquad (4.13)$$

with $E^* = W - M + \frac{1}{2}i\Gamma$. In other words one takes account of the finiteness of the time interval between the creation of the final state at $t = 0$ and the creation of the resonant state at $-t_0 < 0$, having also accounted for the damping induced by the resonance itself.

In analogy with eqs. (4.3), (4.4) one can now formally introduce the quantity Λ_R given by

$$\Lambda_R = \int d^4x\, j_\mu^R(x)\, A_\mu(x)\,, \qquad (4.14)$$

as the action relative to the distribution of classical currents $j_\mu^R(k)$ of eq. (4.11), with

$$j_\mu(x) = \frac{1}{(2\pi)^4} \int d^4k\, j_\mu^R(k)\, e^{-ikx}\,.$$

Similarly, one can introduce a new final state vector, obtained by operating with $e^{i\Lambda_R}$ on the usual final state $|f\rangle$. In this case, however, the operator $e^{i\Lambda_R}$ is no longer unitary, because $\Lambda_R^+ \neq \Lambda_R$. One is therefore led to define

$$|f''\rangle = \frac{1}{\sqrt{N}} e^{i\Lambda_R} |f\rangle , \tag{4.15}$$

with

$$N = \langle e^{-i\Lambda_R^+} e^{i\Lambda_R}\rangle . \tag{4.16}$$

One has

$$\langle e^{-i\Lambda_R^+} e^{i\Lambda_R}\rangle = e^{\frac{1}{2}[\Lambda_R^+, \Lambda_R]} \langle e^{-i(\Lambda_R^+ - \Lambda_R)}\rangle , \tag{4.17}$$

with

$$[\Lambda_R^+, \Lambda_R] = \int d^4k\, \delta(k^2)\, \theta(k_0)\, \{j_\mu(k) j_\mu^*(k) - j_\mu(-k) j_\mu^*(-k)\} , \tag{4.18}$$

where for simplicity we have dropped the label R from the current $j_\mu^R(k)$. Similarly, by series expansion of the exponential in eq. (4.17), and after contraction of the electromagnetic field operators, one finds

$$\langle e^{-i(\Lambda_R^+ - \Lambda_R)}\rangle = \exp\{-\tfrac{1}{2}\int d^4k\, \delta(k^2)\, \theta(k_0)\, [j_\mu^*(k) - j_\mu(-k)]\, [j_\mu^*(-k) - j_\mu(k)]\} . \tag{4.19}$$

From the combined equations (4.17)–(4.19) one finally obtains

$$N = \exp\{\tfrac{1}{2}\int d^4k\, \delta(k^2)\, \theta(k_0) [2 j_\mu^*(k) j_\mu(k) - j_\mu^*(-k) j_\mu^*(k) - j_\mu(-k) j_\mu(k)]\} . \tag{4.20}$$

All the integrals in the above equations have a finite domain of integration, corresponding to $0 \leqslant k_0 \leqslant \Delta\omega$, once $j_\mu^R(k)$ is restricted as in eq. (4.5). The global condition (4.10), imposed on the states (4.15), gives an overall factor in analogy to (4.9), which will be discussed in appendix B.

We are now able to define a matrix element $\overline{M_R} = \langle f''|S|i\rangle$ as in eq. (4.6), where the S-matrix now contains the interaction Hamiltonian responsible for the creation of the resonance R. We therefore have

$$\overline{M_R} = \frac{1}{\sqrt{N}} \langle f| e^{-i\Lambda_R^+} S |i\rangle . \tag{4.21}$$

Formally $\overline{M_R}$ is the matrix element of the operator $S' = \exp[-i\Lambda_R^+] S$ between the states $\langle f|$ and $|i\rangle$ which do not include outgoing soft photons: in other words the infinite number of soft photons created by S is destroyed by $\exp[-i\Lambda_R^+]$. In terms of diagrams, some typical lowest order contributions to (4.21) come from the diagrams of fig. 1, where we have indicated with a small circle the action of the classical source.

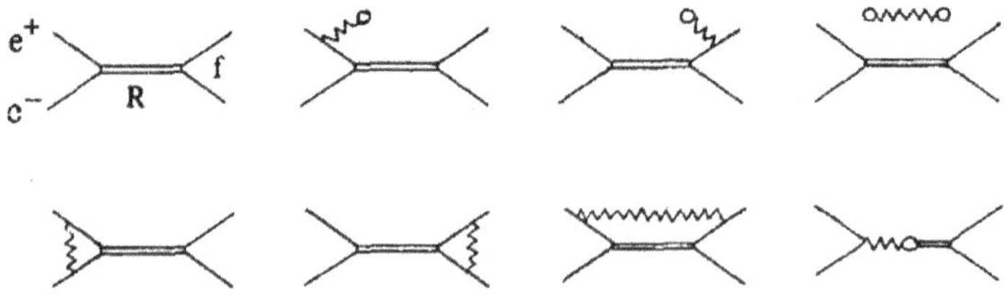

Fig. 1.

Without entering into the details of the calculations, which proceed along the same lines as in ref. [8], we find

$$\bar{M}_R = \frac{(M_R)_\lambda}{\sqrt{N}} \exp\left\{\int_\lambda^{\Delta\omega} \frac{dk}{2k} \int |k|^2 \, d\Omega_k \, [j_\mu^*(k) j_\mu(k) - \tfrac{1}{2} j_\mu^*(k) j_\mu^*(-k)]\right\}, \quad (4.22)$$

where we have dropped, as above, the index R in $j_\mu^R(k)$, and we have consistently introduced, as in eq. (4.6), an auxiliary parameter λ which finally has to go to zero. It should be noticed that in the limit of very large Γ, the resonant factor in eq. (4.11) goes to 1 and we recover all the previous results for the usual classical current.

From eqs. (4.20)–(4.22) one finally obtains

$$\bar{M}_R = (M_R)_\lambda \exp\left\{\tfrac{1}{2} \int_\lambda d^4k \, \delta(k^2) \, \theta(k_0) \, [j_\mu^*(k) j_\mu(k) - i \, \mathrm{Im} \, (j_\mu^*(k) j_\mu^*(-k))]\right\}. \quad (4.23)$$

In order to proceed further we need the explicit expressions for the following integrals:

$$I_1 = \int d^4k \, \delta(k^2) \, \theta(k_0) [j_\mu^*(k) j_\mu(k)] \, , \quad (4.24)$$

$$I_2 = \int d^4k \, \delta(k^2) \, \theta(k_0) [j_\mu^*(k) j_\mu^*(-k)] \, , \quad (4.25)$$

with $\lambda \leqslant k_0 \leqslant \Delta\omega$. From eq. (4.11) one easily finds

$$I_1 = \beta \ln \frac{\Delta\omega}{\lambda} - (\beta_i + 2\beta_{int}) \tfrac{1}{2} \ln \frac{(\Delta\omega - y)^2 + (\frac{1}{2}\Gamma)^2}{y^2 + (\frac{1}{2}\Gamma)^2} + \frac{y\beta_i}{\frac{1}{2}\Gamma} \left[\operatorname{arctg} \frac{\Delta\omega - y}{\frac{1}{2}\Gamma} + \operatorname{arctg} \frac{y}{\frac{1}{2}\Gamma}\right], \quad (4.26)$$

$$I_2 = \beta \ln \frac{\Delta\omega}{\lambda} - (\beta_i + 2\beta_{int}) \tfrac{1}{2} \ln \frac{(y - \frac{1}{2}i\Gamma)^2 - \Delta\omega^2}{(y - \frac{1}{2}i\Gamma)^2} \, , \quad (4.27)$$

where $y = W - M$ and $\beta = \beta_i + 2\beta_{int} + \beta_f$.

Substituting eqs. (4.26) and (4.27) into (4.23) one gets

$$\bar{M}_{\mathrm{R}} = (M_{\mathrm{R}})_\lambda \left(\frac{\Delta\omega}{\lambda}\right)^{\beta/2} \left\{\frac{y^2 + (\frac{1}{2}\Gamma)^2}{(\Delta\omega - y)^2 + (\frac{1}{2}\Gamma)^2}\right\}^{(\beta_{\mathrm{i}} + 2\beta_{\mathrm{int}})/4}$$

$$\times \exp\left\{\frac{y\beta_{\mathrm{i}}}{\Gamma}\left[\operatorname{arctg}\frac{\Delta\omega - y}{\frac{1}{2}\Gamma} + \operatorname{arctg}\frac{y}{\frac{1}{2}\Gamma}\right]\right\} \exp\left\{-\tfrac{1}{4}i(\beta_{\mathrm{i}} + 2\beta_{\mathrm{int}})\right.$$

$$\left.\times\left[\operatorname{arctg}\frac{\Gamma y}{(\frac{1}{2}\Gamma)^2 - y^2} - \operatorname{arctg}\frac{\Gamma y}{\Delta\omega^2 - y^2 + (\frac{1}{2}\Gamma)^2}\right]\right\}. \tag{4.28}$$

As in eq. (4.6), the apparent divergence in $(\Delta\omega/\lambda)^{\beta/2}$ is cancelled by an analogous divergence in $(M_{\mathrm{R}})_\lambda$, coming from the infrared virtual contributions. Moreover all the infrared corrections, in the case of a resonant process, are only those contained in eq. (4.28). In fact for fixed y, if we perform the limit $\Gamma \gg \Delta\omega$, all resonant factors in eq. (4.28) go to 1 and we recover the previous result

$$\bar{M}_{\mathrm{R}} = (M_{\mathrm{R}})_\lambda\ (\Delta\omega/\lambda)^{\beta/2}$$

of eq. (4.6), valid in the case of a matrix element which is a smooth function of the energy.

In view of the application to the case of very narrow resonances, as the recently discovered [6] $\psi(3.1)$ and $\psi(3.7)$, let us consider eq. (4.28) in the limit of $\Gamma \ll \Delta\omega$.

To this purpose we have to specify the λ dependence of the virtual photon effects contained in $(M_{\mathrm{R}})_\lambda$. This is discussed in appendix A. From eqs. (4.28) and (A.2), in the limit of $\Gamma \ll \Delta\omega$, we finally have

$$\bar{M}_{\mathrm{R}} \underset{\Delta\omega \gg \Gamma}{=} \frac{1}{\gamma^{\beta_{\mathrm{f}}/2}\Gamma(1 + \frac{1}{2}\beta_{\mathrm{f}})} (M_{\mathrm{R}})_E \left(\frac{\Delta\omega}{E}\right)^{\beta_{\mathrm{f}}/2} \left(\frac{\frac{1}{2}\Gamma}{E\sin\delta_{\mathrm{R}}}\right)^{\beta_{\mathrm{i}}/2}$$

$$\times \exp(-\tfrac{1}{2}\beta_{\mathrm{i}}\,\delta_{\mathrm{R}} \operatorname{ctg}\delta_{\mathrm{R}}) \exp(-\tfrac{1}{2}i\beta_{\mathrm{i}}\,\delta_{\mathrm{R}})\,, \tag{4.29}$$

where we have defined $M_{\mathrm{R}} \propto e^{i\delta_{\mathrm{R}}} \sin\delta_{\mathrm{R}}$, with $\sin\delta_{\mathrm{R}} = \frac{1}{2}\Gamma/\sqrt{y^2 + (\frac{1}{2}\Gamma)^2}$, and $[\gamma^{\beta_{\mathrm{f}}/2}\Gamma(1 + \frac{1}{2}\beta_{\mathrm{f}})]^{-1}$ is the normalization factor discussed in appendix B (eq. B.6). For the radiatively corrected cross section we finally have

$$(\mathrm{d}\sigma_{\mathrm{R}})_{\mathrm{obs}} \underset{\Delta\omega \gg \Gamma}{=} \frac{1}{\gamma^{\beta_{\mathrm{f}}}\Gamma(1 + \beta_{\mathrm{f}})} (\mathrm{d}\sigma_{\mathrm{R}})_0 \left(\frac{\Delta\omega}{E}\right)^{\beta_{\mathrm{f}}} \left(\frac{\frac{1}{2}\Gamma}{E\sin\delta_{\mathrm{R}}}\right)^{\beta_{\mathrm{i}}}$$

$$\times \exp(-\beta_{\mathrm{i}}\delta_{\mathrm{R}} \operatorname{ctg}\delta_{\mathrm{R}})\,(1 + C_{\mathrm{F}}^{\mathrm{res}})\,, \tag{4.30}$$

where the factor $(1 + C_F^{res})$ stands, as usual, for the left-over finite corrections.

Let us compare this result with the analogous one obtained in the preceding section (eq. (3.14). One can see that the factor $1 - \beta_i \delta_R \operatorname{ctg} \delta_R$, responsible for the radiative tail of the resonance, appears in exponentiated form. This indicates that this exponential factor, which is also present in eq. (4.28), and blows up for y very large, cannot be taken literally, and has to be considered only in the first order approximation. The physical origin of this unpleasant behaviour is due to the fact that, for every large values of y, the emission of a photon with energy $\omega \simeq y$ is not anymore a soft process and imposes strong constraints on the emission of further photons. Such a correlation is not embodied in the coherent approach which uses soft photons emitted independently. A more detailed discussion can be found in appendix B.

Eq. (4.30) has therefore to be transformed into

$$(d\sigma_R)_{obs} \underset{\Delta\omega \gg \Gamma}{=} \frac{1}{\gamma^{\beta_f}\Gamma(1+\beta_f)} (d\sigma_R)_0 \left(\frac{\Delta\omega}{E}\right)^{\beta_f} \left(\frac{\frac{1}{2}\Gamma}{E \sin \delta_R}\right)^{\beta_i}$$

$$\times (1 - \beta_i \delta_R \operatorname{ctg} \delta_R)(1 + C_F^{res}) , \tag{4.31}$$

and can now be used without any restrictions on y. The same considerations apply of course to eq. (4.28).

Let us discuss further our result. Apart from the radiative tail, eq. (4.31) differs from the QED result (eq. (4.9)) only in the fact that a "proper" energy loss $(\frac{1}{2}\Gamma/\sin \delta_R)$ substitutes the external quantity $\Delta\omega$, as far as the emission from the initial state is concerned. A very narrow resonance, in other words, fixes by itself the maximum amount of energy which can be lost from the initial state, in order to still allow the resonance to be produced.

Finally, let us discuss the distribution function $dP(\omega)$ of the emitted radiation for resonant processes. In the normal case one has, as is well known [4],

$$dP(\omega) \simeq \left(\frac{\omega}{E}\right)^{\beta} \frac{d\omega}{\omega} . \tag{4.32}$$

In the resonant case, a similar distribution has not been introduced explicitly. From eq. (4.28) however the desired answer can be obtained in integrated form. In fact one has

$$(d\sigma_R)_{obs} = \int_0^{\Delta\omega} \frac{dP_R(\omega)}{d\omega} d\omega \, (d\sigma_R)_0 , \tag{4.33}$$

and therefore from (4.28) one easily finds

$$dP_R(\omega) \simeq \left(\frac{\omega}{E}\right)^{\beta} \left\{ \beta_f + 2\beta_{int} \frac{y(y-\omega) + (\frac{1}{2}\Gamma)^2}{(y-\omega)^2 + (\frac{1}{2}\Gamma)^2} + \beta_i \frac{y^2 + (\frac{1}{2}\Gamma)^2}{(y-\omega)^2 + (\frac{1}{2}\Gamma)^2} \right\} \frac{d\omega}{\omega} , \tag{4.34}$$

where the approximate equality is due to a linearization in β_i, β_{int} and β_f performed in the bracket of the r.h.s. Of course, in the limit of very large Γ, this reduces to eq. (4.32). Furthermore, for $\beta_f = \beta_{int} = 0$, we recover the known result [5,9,15]

$$(\mathrm{d}\sigma_R)_{obs} \simeq \beta_i \int_0^{\Delta\omega} \frac{\mathrm{d}\omega}{\omega} \left(\frac{\omega}{E}\right)^{\beta_i} \mathrm{d}\sigma_R(W-\omega)\,, \tag{4.35}$$

whereas for $\beta_{int} = 0$, we obtain the result derived in ref. [5]

$$(\mathrm{d}\sigma_R)_{obs} \simeq \int_0^{\Delta\omega} \frac{\mathrm{d}\omega}{\omega} \left(\frac{\omega}{E}\right)^{\beta_i+\beta_f} \{\beta_f\, \mathrm{d}\sigma_R(W) + \beta_i\, \mathrm{d}\sigma_R(W-\omega)\}\,. \tag{4.36}$$

4.3. Interference of a resonant term with a pure QED background

We would like to discuss now the radiative effects arising from the interference of a resonant term with a pure QED term. In other words we have in mind processes like $e^+e^- \to e^+e^- (\mu^+, \mu^-, ...)$ in which the reaction can proceed *via* two different dynamics. For these processes, of course, the detailed knowledge of the radiative effects is necessary for those experiments studying for example forward–backward asymmetries, dips, and similar subtle effects.

Form the point of view discussed in this section, the procedure to follow is quite clear. One starts with the sum of two well defined and finite matrix elements, $\bar{M}_{QED}$ and $\bar{M}_R$, as

$$\bar{M} = \bar{M}_{QED} + \bar{M}_R = \langle f|\, e^{-i\Lambda_c} S_{QED}|i\rangle + \frac{1}{\sqrt{N}} \langle f|\, e^{-i\Lambda_R^+} S_R|i\rangle\,. \tag{4.37}$$

The observed cross section is then proportional to $|\bar{M}|^2$. While $|\bar{M}_{QED}|^2 \propto (\mathrm{d}\sigma_{QED})_{obs}$ and $|\bar{M}_R|^2 \propto (\mathrm{d}\sigma_R)_{obs}$, the interference term comes only from that part of $\langle f|e^{-i\Lambda_c}$ which overlaps with $\langle f|\,e^{-i\Lambda_R^+}/\sqrt{N}$. One is therefore led to define a final state $e^{i\Lambda_R}|f\rangle/\sqrt{N}$ also for the pure QED part of the interaction. We introduce therefore a matrix element $\bar{M}_{QED}^{int}$ defined as

$$\bar{M}_{QED}^{int} = \frac{1}{\sqrt{N}} \langle f|\, e^{-i\Lambda_R} S_{QED}|i\rangle\,. \tag{4.38}$$

All the interference effects will then come from Re$\{\bar{M}_{QED}^{int*}\bar{M}_R\}$, as one can see by comparison with perturbation theory.

Proceeding as in the preceding cases, and using the same notation, we find

$$\bar{M}_{QED}^{int} = (M_{QED})_\lambda \int_\lambda \mathrm{d}^4k\, \delta(k^2)\, \theta(k_0) [j_\mu^{R*}(k)\, j_\mu^c(k) - \tfrac{1}{2} j_\mu^R(k)\, j_\mu^{R*}(k)$$
$$- \tfrac{1}{2} i\, \mathrm{Im}\, (j_\mu^{R*}(k)\, j_\mu^{R*}(-k))]\,, \tag{4.39}$$

where $j_\mu^{\mathrm{R}}(k)$ is defined in eq. (4.11) and $j_\mu^{\mathrm{c}}(k) = I_\mu(k) + F_\mu(k)$ is the usual classical current without resonant modifications. The expression $(M_{\mathrm{QED}})_\lambda = \langle \mathrm{f} | S_{\mathrm{QED}} | \mathrm{i} \rangle_\lambda$, has to be identified with that of eq. (4.6), where the index QED was emitted.

For the first integral on the r.h.s. of eq. (4.39) we easily find

$$I_3 = \int \mathrm{d}^4 k\, \delta(k^2)\, \theta(k_0)\, [j_\mu^{\mathrm{R}*}(k)\, j_\mu^{\mathrm{c}}(k)]$$

$$= \beta \ln\left(\frac{\Delta\omega}{\lambda}\right) - (\beta_{\mathrm{i}} + \beta_{\mathrm{int}}) \ln \frac{y - \frac{1}{2} i\Gamma - \Delta\omega}{y - \frac{1}{2} i\Gamma}\,, \tag{4.40}$$

which, together with eqs. (4.24) and (4.25) finally leads to

$$\bar{M}_{\mathrm{QED}}^{\mathrm{int}} = (M_{\mathrm{QED}})_\lambda \left(\frac{\Delta\omega}{\lambda}\right)^{\beta/2} \left\{\frac{y^2 + (\frac{1}{2}\Gamma)^2}{(\Delta\omega - y)^2 + (\frac{1}{2}\Gamma)^2}\right\}^{\beta_{\mathrm{i}}/4}$$

$$\times \exp\left\{-\frac{y\beta_{\mathrm{i}}}{\Gamma}\left[\mathrm{arctg}\, \frac{\Delta\omega - y}{\frac{1}{2}\Gamma} + \mathrm{arctg}\, \frac{y}{\frac{1}{2}\Gamma}\right]\right\} \exp\left\{-\tfrac{1}{4} i(\beta_{\mathrm{i}} + 2\beta_{\mathrm{int}})\right.$$

$$\times \left[\mathrm{arctg}\, \frac{\Gamma y}{(\frac{1}{2}\Gamma)^2 - y^2} - \mathrm{arctg}\, \frac{y\Gamma}{\Delta\omega^2 - y^2 + (\frac{1}{2}\Gamma)^2}\right]$$

$$\left. - i(\beta_{\mathrm{i}} + \beta_{\mathrm{int}})\, \mathrm{arctg}\, \frac{\Delta\omega(\frac{1}{2}\Gamma)}{(\Delta\omega - y)y - (\frac{1}{2}\Gamma)^2}\right\}. \tag{4.41}$$

As usual the infrared divergence is automatically cancelled from $(M_{\mathrm{QED}})_\lambda$ and $(\Delta\omega/\lambda)^{\beta/2}$, introducing a factor $(M_{\mathrm{QED}})_{\mathrm{E}}(\Delta\omega/E)^{\beta/2}$.

Interference effects come now from the real part of the product $\bar{M}_{\mathrm{R}} \bar{M}_{\mathrm{QED}}^{\mathrm{int}*}$, where both quantities are finite. Combining eqs. (4.28) and (4.41) we obtain finally, in the limit of $\Delta\omega \gg \Gamma$,

$$\bar{M}_{\mathrm{R}} \bar{M}_{\mathrm{QED}}^{\mathrm{int}*} = [\gamma^{\beta_{\mathrm{f}} + \beta_{\mathrm{int}}} \Gamma(1 + \beta_{\mathrm{f}} + \beta_{\mathrm{int}})]^{-1}$$

$$\times (M_{\mathrm{R}})_{\mathrm{E}} (M_{\mathrm{QED}})_{\mathrm{E}}^* \left(\frac{\Delta\omega}{E}\right)^{\beta_{\mathrm{f}} + \beta_{\mathrm{int}}} \left(\frac{\frac{1}{2}\Gamma}{E \sin \delta_{\mathrm{R}}}\, \mathrm{e}^{-i\delta_{\mathrm{R}}}\right)^{\beta_{\mathrm{i}}}. \tag{4.42}$$

where we have also introduced the normalization factor discussed in appendix B. This result again explicitly shows the appearance of a factor $\frac{1}{2}\Gamma/\sin \delta_{\mathrm{R}}\, \mathrm{e}^{i\delta_{\mathrm{R}}}$ as the "proper" energy loss from the initial state, playing the symmetrical role of $\Delta\omega$ for the final state. Notice also that the exponential factor giving rise to the resonance tail, common to both eqs. (4.28) and (4.41), has completely disappeared.

With the standard definition $M_{\mathrm{R}} \propto \sin \delta_{\mathrm{R}}\, \mathrm{e}^{i\delta_{\mathrm{R}}}$, and taking the real part of eq. (4.42) we finally obtain for the interference cross section

$$(d\sigma^{\rm int})_{\rm obs} = \frac{1}{\gamma^{\beta_f+\beta_{\rm int}}\Gamma(1+\beta_f+\beta_{\rm int})}(d\sigma^{\rm int})_0\left(\frac{\Delta\omega}{E}\right)^{\beta_f+\beta_{\rm int}}$$

$$\times\left(\frac{\frac{1}{2}\Gamma}{E\sin\delta_R}\right)^{\beta_i}\{\cos(\delta_R\beta_i)+{\rm tg}\,\delta_R\,\sin(\delta_R\beta_i)\}\,(1+C_F^{\rm int})\,, \qquad (4.43)$$

where $(d\sigma^{\rm int})_0 \propto 2{\rm Re}\,[(M_{\rm QED})_0(M_R)_0^*]$ and $(1+C_F^{\rm int})$ is, as usual, the left-over finite correction. Eq. (4.43) coincides with the result obtained in the preceding section (eq. (3.18)), in the limit $\beta_{\rm int}=0$.

Let us summarize the results of this section. By extending the formalism of ref. [8] to the case of resonance production, we have shown that the introduction of coherent states provides a very powerful tool for an intuitive and physically significant description of the infrared problems also when the presence of a very narrow resonance introduces further subtleties to the problem. In particular all the results can be rigorously proved to all orders in α, without any approximation. They explicitly show that in the presence of a narrow resonance, a physical parameter, the width Γ, enters in the description of the radiative effects.

These results can be also easily extended to the case of a general reaction $a+b\to c+d+...$, proceeding through the formation of a resonance.

5. Phenomenological applications to $\psi(3.1)$ and $\psi'(3.7)$

Some applications of our formulae to the $\psi(3.1)$ and $\psi'(3.7)$ resonances have already been presented in ref. [7]. Here we wish to comment upon several aspects of those applications, and discuss some further ones.

As is well known, since the resonances of interest are very narrow, one has to integrate over the machine resolution, which is assumed to be

$$G(W'-W)=\frac{1}{\sqrt{2\pi}\,\sigma}\,e^{-(W'-W)^2/2\sigma^2}\,, \qquad (5.1)$$

where σ is the machine dispersion, such that $(\Delta W)_{\rm FWHM}=2.3548\,\sigma$.

In the case of hadron production, $\beta_{\rm int}$ and β_f are negligible and therefore one has, for the experimental observed cross section

$$\tilde\sigma(W)=\int G(W'-W)\,dW'\,\sigma(W')\,, \qquad (5.2)$$

where

$$\sigma(W')=\frac{12\pi}{W'^2}\frac{\Gamma_e\Gamma_h}{\Gamma^2}\sin^2\delta_R(W')\left\{\frac{\Gamma}{W'\sin\delta_R(W')}\right\}^{\beta_i}$$

$$\times\{1-\beta_i\delta_R\,{\rm ctg}\,\delta_R\}\{1+C_F^{\rm res}\}\,, \qquad (5.3)$$

with

$$C_F^{\mathrm{res}} = \tfrac{13}{12}\beta_i + \frac{\alpha}{\pi}\left(\tfrac{1}{3}\pi^2 - \tfrac{17}{18}\right).$$

Using the above formulae, fits to both Frascati and SPEAR data [6] have already been presented in ref. [7]. As stated there a useful formula for the observed cross section at the peak is

$$\tilde{\sigma}(M) = \frac{6\pi^2\Gamma_e\Gamma_h}{\sqrt{2\pi}\,\sigma M^2\Gamma}\left(\frac{\Gamma}{M}\right)^{\beta_i} \exp\left(\frac{\Gamma}{2\sqrt{2}\sigma}\right)^2 \left\{\operatorname{ercf}\left(\frac{\Gamma}{2\sqrt{2}\sigma}\right) + \tfrac{1}{2}\beta_i \times E_1\left(\frac{\Gamma^2}{8\sigma^2}\right)\right\}(1 + C_F^{\mathrm{res}}), \tag{5.4}$$

which is a straightforward consequence of (5.2) and (5.3) with only the approximation of setting $[\sin\delta_R(W')]^{-\beta_i} \simeq 1$. In eq. (5.4) the second term in the square bracket represents the contribution from the radiative tail. For resonances whose total width is smaller than the machine resolution, a simple expansion of eq. (5.4) in powers of $(\Gamma/2\sqrt{2}\sigma)$ leads to

$$\tilde{\sigma}(M) = \frac{6\pi^2\Gamma_e\Gamma_h}{\sqrt{2\pi}\,\sigma M^2\Gamma}\left(\frac{\Gamma}{M}\right)^{\beta_i}\left(1 + \frac{\Gamma^2}{8\sigma^2}\right)\left\{1 - \frac{\Gamma}{\sqrt{2\pi}\,\sigma} + \beta_i\left[\ln\frac{2\sqrt{2}\sigma}{\Gamma} - \tfrac{1}{2}\gamma\right]\right\}(1 + C_F^{\mathrm{res}}), \tag{5.5}$$

where $\gamma = 0.5772$ is Euler's constant. As explicitly shown in ref. [7], eq. (5.5) gives an excellent estimate (at the level of 2%) of the parameter $(\Gamma_e\Gamma_h/\Gamma)$ once $\tilde{\sigma}(M)$ is given and the machine dispersion σ is fixed.

In place of eq. (5.5) Yennie [15] obtains (for $\psi(3.1)$) for $\tilde{\sigma}(M)$

$$\tilde{\sigma}(M) \simeq \frac{6\pi^2}{\sqrt{2\pi}\,\sigma}\,\frac{\Gamma_e\Gamma_h}{M^2\Gamma}\left(\frac{2\sqrt{2}\sigma}{M}\right)^{\beta_i}(1 + 0.79\,\beta_i). \tag{5.6}$$

While analytically the two expressions disagree, the numerical estimates for $(\Gamma_e\Gamma_h/\Gamma)$ agree to about 5%, in the case of $\psi(3.1)$. The reliability of our expression (5.4) can be also checked in the completely opposite limit of a resonance whose width is large compared to the energy resolution, e.g. resonances like the ρ or the ω. In this case, by performing the limit $2\sqrt{2}\,\sigma/\Gamma \ll 1$, we obtain from (5.4) the well known result

$$\tilde{\sigma}(M) \simeq \frac{2}{\pi\Gamma}\left(\frac{\Gamma}{M}\right)^{\beta_i}\frac{6\pi^2\Gamma_e\Gamma_h}{M^2\Gamma}(1 + C_F^{\mathrm{res}}).$$

For $\psi(3.1)$ taking the published value [6] from SPEAR $\tilde{\sigma}(M) \simeq 2.3\ \mu$b and $\sigma = 1$ MeV we obtain

$$\frac{\Gamma_e \Gamma_h}{\Gamma} \simeq 4.0 \pm 0.2 \text{ keV} , \tag{5.7}$$

which is in good agreement with our analysis [7] of the Frascati data [16].

A similar analysis for $\psi(3.7)$ gives

$$\frac{\Gamma_e \Gamma_h}{\Gamma} \simeq 2.8 \pm 0.3 \text{ keV} , \tag{5.8}$$

having used $\tilde{\sigma}(M) \simeq 0.75\ \mu$b [6] and $\sigma = 1.15$ MeV ($\Gamma = 1$ MeV) or $\sigma \simeq 1.4$ MeV ($\Gamma = 0.2$ MeV). It is obvious from our formulae that if the peak values are changed, $\Gamma_e \Gamma_h / \Gamma$ changes almost linearly, for a given σ.

Regarding the leptonic modes, the extraction of the dynamical parameters is more complicated since the details of the experimental setups have to be taken into account. As shown in our earlier work [7] no dips due to interference can arise (upon machine integration) in $e^+e^- \to e^+e^-$. Also no significant charge asymmetry results in $e^+e^- \to \mu^+\mu^-$.

It is a pleasure to thank our experimental colleagues at Frascati for discussing all the details of the experimental results. Thanks are also due to Prof. B. Touschek for encouragement and discussions. We have also benefitted from several conversations about the virtual corrections with Profs. C. Altarelli, K. Ellis and R. Petronzio.

Appendix A

Discussion of virtual photon effects

We discuss briefly the infrared properties of the amplitude M_λ defined in the text.

For a smooth amplitude, e.g. that for a pure QED process $e^+e^- \to \mu^+\mu^-$, the infrared dependence of M_λ is well-known,

$$M_\lambda = \left(\frac{\lambda}{E}\right)^{\beta/2} M_E = \left(\frac{\lambda}{E}\right)^{\beta/2} M_0 \sqrt{1 + C_F} , \tag{A.1}$$

where M_0 is the lowest order matrix element and C_F denotes finite contributions (vertex corrections, vacuum polarization, ...), which are to be computed perturbatively, while E is defined as $\frac{1}{2}\sqrt{s}$ and λ is the minimum energy cut-off. If one wishes to perform this calculation in terms of a photon mass λ', the conversion rules may be found from ref. [3].

Notice however that λ^2 does not always scale with s. For example in electron-electron elastic scattering in certain expressions λ^2 scales with t, where t is the momentum transfer [14].

Now we turn to the case of interest, i.e. e^+e^- annihilation through a resonance. In this case one finds that as far as β_i and β_f parts are concerned, λ^2 still scales with s whereas the β_{int} part scales with the resonant propagator. Explicitly,

$$(M_R)_\lambda = (M_R)_E \left(\frac{\lambda}{E}\right)^{(\beta_i+\beta_f)/2} \left\{\frac{\lambda}{\sqrt{y^2+(\frac{1}{2}\Gamma)^2}}\, e^{i\delta R}\right\}^{\beta_{int}} , \tag{A.2}$$

where again $(M_R)_E$ differs from the pure Breit–Wigner (M_R) by finite virtual corrections.

We were led to the above form for the β_{int} dependence upon being informed of explicit second order calculations by Altarelli, Ellis and Petronzio [13].

In the limit of large $\Delta\omega$ (valid for narrow resonances) the β_{int} dependence in (A.2) exactly cancels its counterpart from the real photon emission, as given in the text. This demonstration proceeds the following recognition.

Following Yennie, Frautschi and Suura [3], we define

$$B_{int} = \frac{-ie^2}{(2\pi)^3}\int \frac{d^4k}{k^2-\lambda'^2}\left\{\frac{(2p_1-k)_\mu}{2p_i\cdot k-k^2} - \frac{(2p_2-k)_\mu}{2p_2\cdot k-k^2}\right\}$$

$$\times \left\{\frac{(2p_3-k)_\mu}{2p_3\cdot k-k^2} - \frac{(2p_4-k)_\mu}{2p_4\cdot k-k^2}\right\} S(k) , \tag{A.3}$$

$$\widetilde{B}_{int} = -\pi \int_{k_0<\Delta\omega} \frac{d^3k}{\sqrt{k^2+\lambda'^2}}\, j_\mu^{(i)} j_\mu^{(f)*} S(k) , \tag{A.4}$$

where B_{int} and $\widetilde{B}_{int}$ denote the virtual and real soft photon contributions, while $j_\mu^{(i)}$ and $j_\mu^{(f)}$ are the classical currents defined in the text [eq. (4.5)] and $S(k)$ is the resonant factor [eq. (4.11)]. As shown in ref. [3] the infrared contributions in B_{int} arise from the δ-function part of the photon propagator,

$$\frac{1}{k^2-\lambda'^2+i\epsilon} = \mathrm{PV}\,\frac{1}{k^2-\lambda'^2} - i\pi\delta(k^2-\lambda'^2) .$$

Then a straightforward calculation shows that the infrared factors of B_{int} and $\widetilde{B}_{int}$ cancel exactly in the limit of large $\Delta\omega$ ($\Delta\omega \gg \Gamma$). This argument can be generalized to all orders. Hence the exponentiated form of the virtual infrared factors is completely determined, (A.2), since the corresponding real photon contributions have been explicitly obtained in the text.

The above arguments do not apply to the β_i term, the reason being that the expression similar to (A.3) does not have the resonant factor $S(k)$, while the corresponding (A.4) does. Thus, the infrared singularity, as $\lambda' \to 0$, is indeed cancelled (because $S(k) \to 1$ as $k \to 0$), but the scale of the virtual contributions is provided by the total energy $\sqrt{s}$.

Appendix B

Discussion of the energy conservation constraints

We discuss here the implications coming from the exact conservation of energy in the coherent state method in order to obtain the correct normalization factors as well as to show the correct behaviour of the factor responsible for the radiative tail of the resonance, which in eqs. (4.28)–(4.30) appears in exponentiated form.

The exact procedure of imposing the global condition (4.10) on the final states has been discussed in ref. [8]. We shall adopt here a slightly different, but completely equivalent procedure, by operating on the various matrix elements.

For the classical current, with no resonance present, one has [see eq. (4.6)]

$$\bar{M} = \sum_{n=0}^{\infty} \left\{ \frac{1}{2} \int_{\lambda}^{\Delta\omega} \frac{d^3 k_1}{2k_1} [j_\mu(k_1) j_\mu^*(k_1)] \cdots \int_{\lambda}^{\Delta\omega} \frac{d^3 k_n}{2k_n} \times [(j_\mu(k_n) j_\mu^*(k_n)] \right\}^n \langle f|S|i\rangle_\lambda , \tag{B.1}$$

i.e. $k_{1,\ldots,n} \leqslant \Delta\omega$, when the energy is not fully conserved. Energy conservation can be achieved by introducing a factor

$$I(\Delta\omega) = \begin{cases} 1 & \text{for} \sum k_n \leqslant \Delta\omega \\ 0 & \text{for} \sum k_n > \Delta\omega , \end{cases}$$

which leads to

$$\bar{M}^{\text{cons}} = \frac{1}{2\pi} \int_{-\Delta\omega}^{\Delta\omega} dx \int_{-\infty}^{\infty} e^{-i\sigma x} d\sigma \exp\left\{ \frac{1}{2} \int_{\lambda}^{\infty} \frac{d^3 k}{2k} [j_\mu(k) j_\mu^*(k)] e^{i\sigma k} \right\} \langle f|S|i\rangle_\lambda$$

$$= \frac{1}{\pi} \int_{-\infty}^{\infty} \frac{d\sigma}{\sigma} \sin(\sigma\Delta\omega) \exp\left\{ \frac{1}{2}\beta \int_{\lambda}^{\infty} \frac{dk}{k} e^{i\sigma k} \right\} \langle f|S|i\rangle_\lambda , \tag{B.2}$$

and finally to

$$\bar{M}^{\text{cons}} = \left(\frac{\Delta\omega}{\lambda}\right)^{\beta/2} \frac{1}{\gamma^{\beta/2}} \frac{1}{\Gamma(1+\frac{1}{2}\beta)} M_\lambda$$

$$= \left(\frac{\Delta\omega}{E}\right)^{\beta/2} \frac{1}{\gamma^{\beta/2}} \frac{1}{\Gamma(1+\frac{1}{2}\beta)} M_E , \tag{B.3}$$

with the same notations of eqs. (4.7) and (4.9). Because of the smallness of β one has $[\Gamma(1+\frac{1}{2}\beta)]^2 \simeq \Gamma(1+\beta)$, and therefore a small correction factor $[\gamma^\beta \Gamma(1+\beta)]^{-1}$ which properly normalizes the cross section. As discussed in the text, the smallness of the correction factor reflects the small probability of emitting two or more photons which combine to give a total energy loss $\Delta\omega$.

When there is a resonance present, the same procedure immediately leads to

$$\overline{M}_{\mathrm{R}}^{\mathrm{cons}} = \frac{1}{\pi} \int_{-\infty}^{\infty} \frac{\mathrm{d}\sigma}{\sigma} \sin(\sigma\Delta\omega) \exp\left\{\frac{1}{2} \int_{\lambda}^{\infty} \frac{\mathrm{d}^3 k}{2k} j_\mu(k) j_\mu^*(k)\, \mathrm{e}^{i\sigma k}\right\} (M_{\mathrm{R}})_\lambda \,, \tag{B.4}$$

where $j_\mu(k) \equiv j_\mu^{\mathrm{R}}(k)$, and we have dropped for simplicity the phase factor in eq. (4.23). One has

$$\int_{\lambda}^{\infty} \frac{\mathrm{d}^3 k}{2k} j_\mu(k) j_\mu^*(k)\, \mathrm{e}^{i\sigma k} = \beta E_1(-i\lambda\sigma) - \frac{iy}{\Gamma} \beta_{\mathrm{i}} \{\mathrm{e}^{i\sigma(y+i\Gamma/2)} E_1[i\sigma(y+\tfrac{1}{2}i\Gamma)]$$

$$- \mathrm{e}^{i\sigma(y-i\Gamma/2)} E_1[(i\sigma(y-\tfrac{1}{2}i\Gamma)]\} - \tfrac{1}{2}(\beta_{\mathrm{i}} + 2\beta_{\mathrm{int}})$$

$$\times \{\mathrm{e}^{i\sigma(y+i\Gamma/2)} E_1[i\sigma(y+\tfrac{1}{2}i\Gamma)] + \mathrm{e}^{i\sigma(y-i\Gamma/2)} E_1[i\sigma(y-\tfrac{1}{2}i\Gamma)]\} \,. \tag{B.5}$$

In view of the complexity of eq. (B.5), it is not possible to perform analytically the extra integration in eq. (B.4), for any $\Delta\omega$, Γ and y. However, for very narrow resonances, a compact answer is fortunately achieved by expanding the r.h.s. of (B.5) in terms of $(y \pm \frac{1}{2}i\Gamma)/\Delta\omega$.

One finally obtains

$$\overline{M}_{\mathrm{R}}^{\mathrm{cons}} = \frac{1}{\gamma^{\beta_{\mathrm{f}}/2}} \frac{1}{\Gamma(1+\frac{1}{2}\beta_{\mathrm{f}})} \left(\frac{\Delta\omega}{\lambda}\right)^{\beta_{\mathrm{f}}/2} \left\{\frac{\sqrt{y^2+(\frac{1}{2}\Gamma)^2}}{\lambda}\, \mathrm{e}^{-i\delta_{\mathrm{R}}}\right\}^{\beta_{\mathrm{i}}/2+\beta_{\mathrm{int}}}$$

$$\times \exp(-\tfrac{1}{2}\beta_{\mathrm{i}}\delta_{\mathrm{R}} \operatorname{ctg} \delta_{\mathrm{R}}) (M_{\mathrm{R}})_\lambda \,, \tag{B.6}$$

which coincides with eq. (4.29) apart from the normalization factor $\gamma^{\beta_{\mathrm{f}}/2}\Gamma(1+\frac{1}{2}\beta_{\mathrm{f}})$.

In the same limit of a very narrow resonance, one obtains a similar result for $(\overline{M}_{\mathrm{QED}}^{\mathrm{int}})$. Explicitly we find

$$(\overline{M}_{\mathrm{QED}}^{\mathrm{int}})^{\mathrm{cons}} = \frac{1}{\gamma^{\beta_{\mathrm{f}}/2+\beta_{\mathrm{int}}}} \frac{1}{\Gamma(1+\frac{1}{2}\beta_{\mathrm{f}}+\beta_{\mathrm{int}})} \left(\frac{\Delta\omega}{\lambda}\right)^{\beta_{\mathrm{f}}/2+\beta_{\mathrm{int}}}$$

$$\times \left\{\frac{\sqrt{y^2+(\frac{1}{2}\Gamma)^2}}{\lambda}\, \mathrm{e}^{i\delta_{\mathrm{R}}}\right\}^{\beta_{\mathrm{i}}/2} \exp(\tfrac{1}{2}\beta_{\mathrm{i}}\, \delta_{\mathrm{R}} \operatorname{ctg} \delta_{\mathrm{R}}) (M_{\mathrm{QED}})_\lambda \,. \tag{B.7}$$

From the above results it therefore follows that the normalization factors depend upon the same coefficients which appear as powers of $\Delta\omega$. This is very satisfactory from a physical point of view, because of the probability meaning associated with those factors.

Finally let us turn to the problem of the factor responsible for the radiative tail of the resonance, which appears in eqs. (B.6) and (4.28)–(4.30) in exponentiated form, which in the perturbative approach [see eq. (3.14)] has the correct $1/y$ behaviour. It is easy to convince ourselves that this feature only depends upon the expansion used in the r.h.s. of eq. (B.5), to obtain eq. (B.6). By performing the limit y in $y \gg \Delta\omega$ in (B.5), for example, the correct $1/y$ behaviour can be shown to arise naturally.

References

[1] J.M. Jauch and F. Rohrlich, The theory of photons and electrons (Addison-Wesley, Cambridge, Mass., 1955).
[2] D.R. Yennie and H. Suura, Phys. Rev. 105 (1957) 1378.
[3] D.R. Yennie, S.C. Frautschi and H. Suura, Ann. of Phys. 13 (1961) 379.
[4] E. Etim, G. Pancheri and B. Touschek, Nuovo Cimento 51B (1967) 276.
[5] G. Pancheri, Nuovo Cimento 60 (1969) 321.
[6] J.J. Aubert et al., Phys. Rev. Letters 33 (1974) 1404;
J.E. Augustin et al., Phys. Rev. Letters 33 (1974) 1406, 1453;
C. Bacci et al., Phys. Rev. Letters 33 (1974) 1408;
DASP Collaboration, Phys. Letters 53B (1974) 393;
L. Criegee et al., Phys. Letters 53B (1975) 489
[7] M. Greco, G. Pancheri-Srivastava and Y.N. Srivastava, Frascati Internal report LNF-75/9; Phys. Letters 56B (1975) 367.
[8] M. Greco and G. Rossi, Nuovo Cimento 50 (1967) 168.
[9] G. Bonneau and F. Martin, Nucl. Phys. B27 (1971) 381.
[10] I.B. Khriplovich, Novosibirsk preprint, submitted to the Chicago-Batavia Conf., 1972.
[11] R.W. Brown, V.K. Cung, K.O. Mikaelian and E.A. Pashcos, Phys. Letters 43B (1973) 403.
[12] F.A. Berends, K.J.F. Gaemers and R. Gastmans, Nucl. Phys. B68 (1974) 541; B63 (1973) 381.
[13] G. Altarelli, R.K. Ellis and A. Petronzio, Rome preprint (1975).
[14] Y.S. Tsai, Phys. Rev. 120 (1960) 269.
[15] D.R. Yennie, Phys. Rev. Letters 34 (1975) 239.
[16] R. Baldini-Celio et al., Nuovo Cimento Letters 11 (1974) 711.

Nuclear Physics B171 (1980) 118–140

RADIATIVE CORRECTIONS TO $e^+e^- \to \mu^+\mu^-$ AROUND THE Z_0

M. GRECO

INFN-Laboratori Nazionali di Frascati, Frascati, Italy

G. PANCHERI-SRIVASTAVA[1] and Y. SRIVASTAVA[1]

Northeastern University, Boston, Mass., USA

Received 3 March 1980

Analytical expressions are presented for e.m. radiative effects in the reaction $e^-e^+ \to \mu^-\mu^+$, in the vicinity of the Z_0 mass, in the Weinberg-Salam model. Our formulae contain all finite first-order corrections, as well as soft photon effects resummed to all orders, with no restriction on the relative magnitudes of the energy resolution $\Delta\omega$ and the Z_0 width Γ. Hard photon effects, other than the tail effect, are not included. Weak interactions are considered only to lowest order.

Radiative corrections are explicitly shown to change the lowest-order results drastically, and must therefore be taken in full account for a realistic description of experiments at LEP energies.

1. Introduction

High-energy e^-e^+ colliding beams provide a unique opportunity [1] for testing the properties of weak neutral currents, particularly for energies near the intermediate boson mass. Interference with electromagnetic processes gives characteristic effects which are extremely sensitive to the weak vector and axial couplings.

Experimental investigation of these effects will provide, however, a sensitive test of the theory [2] only when radiative corrections have been fully taken into account. More precisely, we can distinguish two types of radiative corrections, namely those associated with higher-order pure weak effects and the more standard ones due to electromagnetic effects.

The radiative corrections of the first kind, which are finite by virtue of the renormalizability [3] of the theory, themselves play an important role as a test of the theory. They have been recently calculated by Passarino and Veltman [4] for $e^-e^+ \to \mu^-\mu^+$ and by Consoli [5] for $e^-e^+ \to e^-e^+$.

On the other hand, pure electromagnetic radiative corrections, which add no additional theoretical information, are expected to be rather important, especially

[1] Work supported in part by the National Science Foundation, USA.

in the vicinity of the Z_0 boson, and could sizably change the naive expectations. In particular, one must evaluate the corrections due to soft photon emission, which become increasingly important as the energy increases, to all orders in α. Usually, i.e., for non-resonant cross sections, this leads to the exponentiated form $(\Delta\omega/E)^{(4\alpha/\pi)\ln(2E/m)}$, where $\Delta\omega$ is the resolution and $2E$ is the c.m. energy of the experiment. In the case of resonance production, the above factor is modified and for a very narrow resonance (like J/ψ) the correction [6, 7] becomes $\propto(\Gamma/M)^{(4\alpha/\pi)\ln(2E/m)}$. Physically, this is understood by saying that the width Γ provides a natural cut-off in damping the energy loss in the initial state. For the case of the Z_0 boson, where neither of the preceding cases applies $(\Gamma/M\sim\Delta\omega/E)$, the soft correction would be a complicated function of E, M, $\Delta\omega$ and Γ raised to the power $(4\alpha/\pi)\ln(2E/m)$, of the order of 50%.

Another important electromagnetic effect is the presence of a substantial radiative tail. Due to the emission of a hard photon from the initial state, the radiative tail is expected to radically modify the angular asymmetries in $e^+e^-\to\mu^-\mu^+$. It follows, therefore, that a detailed calculation of the e.m. radiative corrections is of primary importance for the forthcoming experiments around the Z_0 mass.

In this paper we present a detailed study of the above electromagnetic effects in the reaction $e^+e^-\to Z_0\to\mu^+\mu^-$ and give simple analytical prescriptions of immediate experimental application. A brief account of our results has already been given [8].

We consider the infrared factors to all orders in perturbation theory. This is achieved by using the coherent state formalism developed previously in ref. [7] for a purely resonant cross section as well as for the interference with a QED background. Simple expressions for the infrared factors are thus obtained, which depend on E, M, Γ and $\Delta\omega$. Finite corrections of order α have also been computed, for the case of unpolarized as well as transversely polarized $e^\pm$ beams. The weak effects are considered only to lowest order, the weak boson Z_0 being taken into account as a resonance of mass M and width Γ in the s-channel. Hard photon effects, other than the tail effect, are not included.

In sect. 2 we define our notation and collect the relevant formulae. The infrared corrections are discussed in detail in sect. 3, while a careful analysis of diagrams contributing to first order in α is presented in sect. 4. A discussion of our results as well as some numerical applications are finally presented in sect. 5.

2. Notation and formulae

For the reaction

$$e^-(p_1)+e^+(p_2)\to\mu^-(p_3)+\mu^+(p_4),$$

we have

$$s = W^2 = 4E^2 = (p_1 + p_2)^2, \tag{2.1a}$$

$$z = \cos\theta = \hat{p}_1 \cdot \hat{p}_3 = \hat{p}_2 \cdot \hat{p}_4 , \tag{2.1b}$$

$$a = \sin\tfrac{1}{2}\theta, \qquad b = \cos\tfrac{1}{2}\theta , \tag{2.1c}$$

$$\beta_e = \frac{4\alpha}{\pi}\left[\ln\frac{W}{m_e} - \frac{1}{2}\right], \tag{2.2a}$$

$$\beta_\mu = \frac{4\alpha}{\pi}\left[\ln\frac{W}{m_\mu} - \frac{1}{2}\right], \tag{2.2b}$$

$$\beta_{\text{int}} = \frac{4\alpha}{\pi}\ln\left(\tan\tfrac{1}{2}\theta\right), \tag{2.2c}$$

$$\Delta \equiv \frac{\Delta\omega}{E} = \text{fractional energy resolution}.$$

The weak boson Z_0 is taken to be a resonance of mass M and total width Γ, such that the phase shift δ_R is

$$\tan\,\delta_R(s) = \frac{M\Gamma}{M^2 - s} . \tag{2.3}$$

In the standard Weinberg-Salam model [2], with only one doublet of Higgs fields one has

$$M = \left(\frac{\pi\alpha}{\sqrt{2}\,G_F}\right)^{1/2} \frac{1}{\sin\theta_W \cos\theta_W} , \tag{2.4}$$

$$\Gamma \equiv \Gamma(Z_0 \to \text{all}) \simeq \frac{G_F M^3}{24\sqrt{2}\,\pi}\left\{\left[1 + \left(1 - 4\sin^2\theta_W\right)^2\right]N_l + 2N_\nu\right.$$

$$\left. + 3\left[1 + \left(1 - \tfrac{8}{3}\sin^2\theta_W\right)^2\right]N_u + 3\left[1 + \left(1 - \tfrac{4}{3}\sin^2\theta_W\right)^2\right]N_d\right\}, \tag{2.5}$$

where N_l, N_ν, N_u and N_d are the numbers of charged leptons, neutrinos, charge $\frac{2}{3}$ and $-\frac{1}{3}$ quarks respectively, with masses appreciably smaller than $\frac{1}{2}M$. Using $\sin^2\theta_W \simeq \frac{1}{4}$ and $N_l = N_\nu = N_u = N_d = 3$, eqs. (2.4), (2.5) give $M \simeq 86.4$ GeV, $\Gamma \simeq 2.2$ GeV and $\Gamma_e' \equiv \Gamma(Z_0 \to e\bar{e}) = \frac{1}{32}\Gamma$.

We also define

$$r_V = \frac{g_V^2}{g_V^2 + g_A^2} = \frac{(1 - 4\sin^2\theta_W)^2}{1 + (1 - \sin^2\theta_W)^2}, \tag{2.6a}$$

$$r_A = \frac{g_A^2}{g_V^2 + g_A^2} = \frac{1}{1 + (1 - 4\sin^2\theta_W)^2}, \tag{2.6b}$$

$$r_{AV} = \frac{g_A g_V}{g_V^2 + g_A^2} = \frac{4\sin^2\theta_W - 1}{1 + (1 - 4\sin^2\theta_W)^2}, \tag{2.6c}$$

with $g_V = (4\sin^2\theta_W - 1)/2\sin(2\theta_W)$, $g_A = \frac{1}{2}\sin(2\theta_W)$ and

$$\Gamma_e = \tfrac{1}{3}\alpha M(g_V^2 + g_A^2), \tag{2.7a}$$

$$\chi(s) = \left(\frac{3\Gamma_e}{\alpha M}\right)\left(\frac{s}{s - M^2 + iM\Gamma}\right). \tag{2.7b}$$

The differential cross section can be written as follows:

$$\begin{aligned}\left(\frac{d\sigma}{d\Omega}\right)^{corr} &= C_{infra}^{res}\left(\frac{d\sigma_{res}}{d\Omega}\right)(1 + C_F^{res}) + C_{infra}^{int,V}\left(\frac{d\sigma_{int,V}}{d\Omega}\right)\\ &\times(1 + C_F^{int,V}) + C_{infra}^{int,A}\left(\frac{d\sigma_{int,A}}{d\Omega}\right)(1 + C_F^{int,A})\\ &+ C_{infra}^{int,VA}\left(\frac{d\sigma_{int,VA}}{d\Omega}\right)(1 + C_F^{int,VA})\\ &+ C_{infra}^{QED}\left(\frac{d\sigma_{QED}}{d\Omega}\right)(1 + C_F^{QED}),\end{aligned} \tag{2.8}$$

where the Born cross sections are given by*

$$\left(\frac{d\sigma_{QED}}{d\Omega}\right) = \frac{\alpha^2}{4s}(1 + z^2 - P(\theta, \phi)), \tag{2.9a}$$

$$\left(\frac{d\sigma_{int,V}}{d\Omega}\right) = \frac{\alpha^2}{4s}(1 + z^2 - P(\theta, \phi))\cdot(2\,\mathrm{Re}\,\chi)r_V, \tag{2.9b}$$

$$\left(\frac{d\sigma_{int,B}}{d\Omega}\right) = \frac{\alpha^2}{4s}(2z)(2\,\mathrm{Re}\,\chi)r_A, \tag{2.9c}$$

* See, for example, ref. [9] for the general case of axial and vector couplings, as well as in the presence of longitudinal polarization of the e^+e^- beams.

$$\left(\frac{d\sigma_{\text{int,VA}}}{d\Omega}\right)=\frac{\alpha^2}{4s}Q(\theta,\phi)(2\,\text{Im}\,\chi)r_{\text{VA}},\tag{2.9d}$$

$$\left(\frac{d\sigma_{\text{res}}}{d\Omega}\right)=\frac{\alpha^2}{4s}\{1+z^2+8zr_{\text{A}}r_{\text{V}}-P(\theta,\phi)\cdot(r_{\text{V}}^2-r_{\text{A}}^2)\}|\chi|^2,\tag{2.9e}$$

with $P(\theta,\phi)\equiv|\xi_+\xi_-|\sin^2\theta\cos\phi$ and $Q(\theta,\phi)\equiv|\xi_+\xi_-|\sin^2\theta\sin 2\phi$.

In the above $\xi_{(\pm)}$ is the polarization of the $e^{(\pm)}$ beam along the direction of the magnetic field and (θ,ϕ) refer to the μ^- with respect to the e^-.

The $C_{\text{infra}}^{\text{res, int, QED}}$ are the infrared factors associated with each of the respective cross sections, while $C_{\text{F}}^{(i)}$ ($i=$ res, int(V, A, VA), QED) incorporate the rest of the finite corrections. The C_{infra} factors are obtained using the techniques developed in ref. [7] for narrow resonances, and are discussed in detail in sect. 3. They take the form

$$C_{\text{infra}}^{\text{QED}}=\left(\frac{\Delta\omega}{E}\right)^{\beta_e+\beta_\mu+2\beta_{\text{int}}},\tag{2.10a}$$

$$C_{\text{infra}}^{\text{int},(i)}=\left(\frac{\Delta\omega}{E}\right)^{\beta_\mu+\beta_{\text{int}}}\frac{1}{\cos\delta_{\text{R}}}$$

$$\times\text{Re}\left\{e^{i\delta_{\text{R}}}\left(\frac{\Delta}{1+\Delta(s/M\Gamma)e^{i\delta_{\text{R}}}\sin\delta_{\text{R}}}\right)^{\beta_e}\left(\frac{\Delta}{\Delta+(M\Gamma/s)e^{-i\delta_{\text{R}}}/\sin\delta_{\text{R}}}\right)^{\beta_{\text{int}}}\right\},$$

$$(i=\text{V},\text{A},\text{VA}),\tag{2.10b}$$

$$C_{\text{infra}}^{\text{res}}=\left(\frac{\Delta\omega}{E}\right)^{\beta_\mu}\left|\frac{\Delta}{1+\Delta(s/M\Gamma)e^{i\delta_{\text{R}}}\sin\delta_{\text{R}}}\right|^{\beta_e}$$

$$\times\left|\frac{\Delta}{\Delta+(M\Gamma/s)\,e^{-i\delta_{\text{R}}}/\sin\delta_{\text{R}}}\right|^{2\beta_{\text{int}}}\left[1-\beta_e\delta(s,\Delta\omega)\cot\delta_{\text{R}}\right],\tag{2.10c}$$

with $\delta(s,\Delta\omega)$ given by eq. (3.8).

Finally, the $C_{\text{F}}^{(i)}$ factors are

$$C_{\text{F}}^{\text{QED}}=\tfrac{13}{12}(\beta_e+\beta_\mu)+\frac{2\alpha}{\pi}\left(\tfrac{1}{3}\pi^2-\tfrac{17}{18}\right)+\delta_{\text{VP}}+X_{\text{V}}(\theta,\phi),\tag{2.11a}$$

$$C_{\text{F}}^{\text{int,V}}=\tfrac{11}{12}(\beta_e+\beta_\mu)+\frac{2\alpha}{\pi}\left(\tfrac{1}{3}\pi^2-\tfrac{13}{18}\right)+\tfrac{1}{2}\delta_{\text{VP}}+\tfrac{1}{2}X_{\text{V}}(\theta,\phi),\tag{2.11b}$$

$$C_{\text{F}}^{\text{int,A}}=\tfrac{11}{12}(\beta_e+\beta_\mu)+\frac{2\alpha}{\pi}\left(\tfrac{1}{3}\pi^2-\tfrac{13}{18}\right)+\tfrac{1}{2}\delta_{\text{VP}}+\tfrac{1}{2}X_{\text{A}}(\theta,\phi),\tag{2.11c}$$

$$C_F^{\text{int,VA}} = \tfrac{11}{12}(\beta_e + \beta_\mu) + \frac{2\alpha}{\pi}\left(\tfrac{1}{3}\pi^2 - \tfrac{13}{18}\right) + \tfrac{1}{2}\delta_{VP} + \tfrac{1}{2}X_{AV}(\theta,\phi), \tag{2.11d}$$

$$C_F^{\text{res}} = \tfrac{3}{4}(\beta_e + \beta_\mu) + \frac{2\alpha}{\pi}\left(\tfrac{1}{3}\pi^2 - \tfrac{1}{2}\right) - \tfrac{2}{3}\alpha\left(\frac{\alpha M^2\Gamma}{\Gamma_e s}\right)$$
$$\times \frac{r_V Y_V(\theta,\phi) + r_A Y_A(\theta,\phi)}{1 + z^2 + 8 z r_V r_A - (r_V^2 - r_A^2)\cdot P(\theta,\phi)}, \tag{2.11e}$$

where

$$X_V(\theta,\phi) = -\frac{\alpha}{\pi}\left\{ z\left(\frac{\ln^2 a}{b^4} + \frac{\ln^2 b}{a^4}\right) - \left(\frac{\ln a}{b^2} - \frac{\ln b}{a^2}\right)\right.$$
$$+ 4\left[\ln^2 b - \ln^2 a + \tfrac{1}{2}\mathrm{Li}_2(a^2) - \tfrac{1}{2}\mathrm{Li}_2(b^2)\right] + \frac{2z}{1 + z^2 - P(\theta,\phi)}$$
$$\left.\times\left[z\left(\frac{\ln^2 a}{b^4} - \frac{\ln^2 b}{a^4}\right) - \left(\frac{\ln a}{b^2} + \frac{\ln b}{a^2}\right)\right]\right\}, \tag{2.12a}$$

$$X_A(\theta,\phi) = -\frac{2\alpha}{\pi}\left\{ \mathrm{Li}_2(a^2) - \mathrm{Li}_2(b^2) - (\ln a)^2 + (\ln b)^2 - \frac{b^2}{z}\ln a - \frac{a^2}{z}\ln b\right.$$
$$\left.- \frac{P(\theta,\phi)}{4z}\cdot\left[z\left(\frac{\ln^2 a}{b^4} - \frac{\ln^2 b}{a^4}\right) - \left(\frac{\ln a}{b^2} + \frac{\ln b}{a^2}\right)\right]\right\}, \tag{2.12b}$$

$$X_{AV}(\theta,\phi) = -\frac{2\alpha}{\pi}\left\{ z\frac{\ln^2 b}{a^4} + \frac{\ln b}{a^2} + 2(\ln^2 b - \ln^2 a) + \mathrm{Li}_2(a^2)\right.$$
$$\left.- \mathrm{Li}_2(b^2) - \pi\frac{\mathrm{Re}\chi}{\mathrm{Im}\chi}\left[z\frac{\ln b}{a^4} - 2\ln\left(\frac{a}{b}\right) + \frac{1}{1-z} + \tfrac{1}{3}\bar{R}\right]\right\}, \tag{2.12c}$$

$$Y_V(\theta,\phi) = \left\{\left[z - 2z(\ln a + \ln b) + 2(1 + z^2)\left(\ln\frac{a}{b} - \tfrac{1}{6}\bar{R}\right)\right]\right.$$
$$\left.- P(\theta,\phi)\cdot\left[\left(2\ln\frac{a}{b} - \tfrac{1}{3}\bar{R}\right) - \tfrac{1}{2}z\left(\frac{\ln a}{b^4} + \frac{\ln b}{a^4}\right) - \frac{z}{1-z^2}\right]\right\}, \tag{2.12d}$$

$$Y_A(\theta,\phi) = \left\{\left[2z\left(\ln\frac{a}{b} - \tfrac{1}{3}\bar{R}\right) + 1\right]\right.$$
$$\left.+ P(\theta,\phi)\cdot\left[\frac{1}{1-z^2} - \tfrac{1}{2}z\left(\frac{\ln a}{b^4} - \frac{\ln b}{a^4}\right)\right]\right\}, \tag{2.12e}$$

with $\bar{R} = \Sigma_{i=\text{leptons+quarks}} Q_i^2$ and δ_{VP} is given in sect. 4 (eq. 4.32).

This completes the list of our formulae. Their derivation and discussion follow in the coming sections.

3. Multiple soft photons effects

In this section we will discuss the collective effect of multiple soft photons emission in the framework of the coherent state formalism, leading to the expressions of the infrared factors given above.

The use of coherent states in the infrared problem in QED is rather well known [10]. This technique was generalized in ref. [7] to discuss the radiative effects generated in presence of a narrow resonance like the J/ψ, whose decay width is much smaller than the energy resolution of the experiment. To our purposes, as well as for the readers' convenience, we will briefly review those results, without any restriction on the relative size of Γ and $\Delta\omega$.

When the reaction

$$a+b\to c+d+\dots, \tag{3.1}$$

proceeds through the production of a resonance in the s-channel, the classic current associated to the external charged particles is modified as follows:

$$j_\mu^{\rm R}(k)=\frac{W-M+\frac{1}{2}i\Gamma}{W-M-k+\frac{1}{2}i\Gamma}j_\mu^{(\rm i)}(k)+j_\mu^{(\rm f)}(k)\,, \tag{3.2}$$

where $j_\mu^{(\rm i)}(k)$ and $j_\mu^{(\rm f)}(k)$ are the usual classical currents associated to the initial and final particles respectively, namely

$$j_\mu^{(\rm i,\,f)}(k)=\frac{ie}{(2\pi)^{3/2}}\sum_{1}^{(\rm i,\,f)}\varepsilon_1\frac{p_\mu^{(1)}}{(p_1\cdot k)}\,. \tag{3.3}$$

This modification takes into account the finiteness of the time interval between the formation of the resonance and the creation of the final state, as can be seen from the Fourier transform of (3.2). In perturbation theory this is equivalent to write the matrix element for the emission of one soft photon in the reaction (3.1) as

$$M_\mu^{(1\gamma)}(k)=j_\mu^{(\rm i)}M^0(W-k)+j_\mu^{(\rm f)}M^0(W)\,, \tag{3.4}$$

and accounts for the shift in the c.m. energy when the radiation is emitted from the initial state.

Let us introduce the action $\Lambda_{\rm R}$ relative to the distribution of classical current $j^{\rm R}(k)$ of eq. (3.2):

$$\Lambda_{\rm R}=\frac{1}{(2\pi)^4}\int{\rm d}^4k\,j_\mu^{\rm R}(k)A^\mu(-k)\,, \tag{3.5}$$

where $A_\mu(k)$ is the quantized electromagnetic field. Then, for a pure resonant process, the matrix element

$$\overline{M}_{\mathrm{R}} = \frac{1}{\sqrt{N}} \langle \mathrm{f} | \mathrm{e}^{-i\Lambda_{\mathrm{R}}^{+}} S | \mathrm{i} \rangle , \tag{3.6}$$

with $N = \langle \mathrm{e}^{-i\Lambda_{\mathrm{R}}^{+}} \mathrm{e}^{i\Lambda_{\mathrm{R}}} \rangle$, can be shown [7] to be: (i) finite and does not therefore possess an infrared divergence, (ii), separable in the infrared factors and (iii) directly comparable with the observable cross section which results proportional to $|\overline{M}_{\mathrm{R}}|^2$.

Without entering into the details of the derivation, one obtains for $\mathrm{e}\bar{\mathrm{e}} \to \mathrm{R} \to \mu^+ \mu^-$

$$\mathrm{d}\sigma_{\mathrm{corr}}^{\mathrm{res}} = \mathrm{d}\sigma_0^{\mathrm{res}} \left(\frac{\Delta\omega}{E} \right)^{\beta_\mu} \left| \frac{2\Delta\omega}{\sqrt{s}} \frac{M_{\mathrm{R}}^2 - s}{M_{\mathrm{R}}^2 - s + 2\sqrt{s}\,\Delta\omega} \right|^{\beta_{\mathrm{e}}}$$

$$\times \left| \frac{2\sqrt{s}\,\Delta\omega}{M_{\mathrm{R}}^2 - s + 2\sqrt{s}\,\Delta\omega} \right|^{2\beta_{\mathrm{int}}} \left\{ 1 + \frac{s - M^2}{M\Gamma} \beta_{\mathrm{e}} \delta(s, \Delta\omega)(1 + C_{\mathrm{F}}^{\mathrm{res}}) \right\}, \tag{3.7}$$

where $M_{\mathrm{R}}^2 = M^2 - iM\Gamma$,

$$\delta(s, \Delta\omega) = \operatorname{arctg} \frac{2\sqrt{s}\,\Delta\omega - (s - M^2)}{M\Gamma} + \operatorname{arctg} \frac{s - M^2}{M\Gamma}, \tag{3.8}$$

and $C_{\mathrm{F}}^{\mathrm{res}}$ accounts for the rest of the finite corrections and has to be calculated perturbatively.

Let us briefly discuss our result. First, it is easily seen that the infrared factors appearing in eq. (3.7) reduce to the standard one $(\Delta\omega/E)^{\beta_{\mathrm{e}} + \beta_\mu + 2\beta_{\mathrm{int}}}$ in the limit $\Delta\omega \ll |(M_{\mathrm{R}}^2 - s)/2\sqrt{s}|$, as they should.

On the other hand, in the case of a narrow resonance like the J/ψ, for which in a typical experiment $\Delta\omega \gg |(M_{\mathrm{R}}^2 - s)/2\sqrt{s}|$, the $\Delta\omega$ dependence drops completely out, namely the width of the resonance provides a natural cut-off in damping the energy loss in the initial state. Furthermore, the β_{int} dependence in eq. (3.8) also cancels out, giving no interference between the soft emission from the initial and final states.

Finally the term proportional to $\delta(s, \Delta\omega)$, with the arctangent defined to have values in the interval $(-\frac{1}{2}\pi, \frac{1}{2}\pi)$, gives the radiative tail of the resonance. In fact for narrow resonances $\delta(s, \Delta\omega)$ reduces to $\delta_{\mathrm{R}}(s)$, the usual phase shift of the Breit-Wigner resonance.

With the standard definition [see eq. (2.3)]

$$\frac{s}{M_{\mathrm{R}}^2 - s} \equiv \frac{s}{M\Gamma} \sin \delta_{\mathrm{R}} \mathrm{e}^{i\delta_{\mathrm{R}}}, \tag{3.9}$$

eq. (3.7) easily leads to the infrared resonant factor given in sect. 2 [eq. (2.10c)].

We would now like to discuss the radiative effect arising from the interference of a resonant term with a pure QED term. As shown in detail in ref. [7], the procedure to follow is quite clear. One starts with the sum of two infrared-finite matrix elements, $\overline{M}_{\rm QED}$ and $\overline{M}_{\rm R}$, as

$$\overline{M} = \overline{M}_{\rm QED} + \overline{M}_{\rm R} = \langle {\rm f}| {\rm e}^{-i\Lambda_{\rm c}} S_{\rm QED}|{\rm i}\rangle + \frac{1}{\sqrt{N}}\langle {\rm f}| {\rm e}^{-i\Lambda_{\rm R}^{+}} S_{\rm R}|{\rm i}\rangle\,, \tag{3.10}$$

where $\Lambda_{\rm c}$ is the action relative to the distribution of pure classical currents

$$j_\mu^{\rm c}(k) = j_\mu^{(\rm i)}(k) + j_\mu^{(\rm f)}(k)\,, \tag{3.11}$$

which describes the infrared properties of a pure QED process [10]. More explicitly $\Lambda_{\rm c}$ is given by eq. (3.5) when $j_\mu^{\rm R}(k) \to j_\mu^{\rm c}(k)$. The full observed cross section is then proportional to $|\overline{M}|^2$. While $|\overline{M}_{\rm QED}|^2 \propto {\rm d}\sigma_{\rm corr}^{\rm QED}$ and $|\overline{M}_{\rm R}|^2 \propto {\rm d}\sigma_{\rm corr}^{\rm res}$, the interference term comes only from that part of $\langle {\rm f}|\exp(-i\Lambda_{\rm c})$ which overlaps with $\langle {\rm f}|\exp(-i\Lambda_{\rm R}^{+})/\sqrt{N}$. Introducing, therefore, a matrix element $\overline{M}_{\rm QED}^{\rm int}$ as

$$\overline{M}_{\rm QED}^{\rm int} = \frac{1}{\sqrt{N}}\langle {\rm f}| {\rm e}^{-i\Lambda_{\rm R}^{+}} S_{\rm QED}|{\rm i}\rangle\,, \tag{3.12}$$

all the interference effects will then come from ${\rm Re}(\overline{M}_{\rm QED}^{\rm int}\overline{M}_{\rm R}^{*})$, as one can see by comparison with perturbation theory.

With that in mind, one finds

$$\begin{aligned} {\rm d}\sigma_{\rm corr}^{\rm int} = {\rm d}\sigma_0^{\rm int}\left(\frac{\Delta\omega}{E}\right)^{\beta_\mu+\beta_{\rm int}} \frac{1}{\cos\delta_{\rm R}} \\ \times {\rm Re}\left\{ {\rm e}^{i\delta_{\rm R}} \left(\frac{\Delta}{1+\Delta s/(M_{\rm R}^2-s)}\right)^{\beta_e} \left(\frac{\Delta}{\Delta + (M_{\rm R}^2-s)/s}\right)^{\beta_{\rm int}}\right\}, \end{aligned} \tag{3.13}$$

which, using eq. (3.9), leads finally to eq. (2.10b).

Again, taking the limit $\Delta \ll (\Delta \gg)|(M_{\rm R}^2 - s)/s|$ one gets the usual QED, or narrow resonance-like behaviour respectively.

This concludes our discussion of the multiple soft photon effect. From our formulae it is clear that, in experiments studying forward-backward asymmetries, dips and similar subtle effects, the Γ and $\Delta\omega$ dependence in the $C_{\rm infra}^{(i)}$ factors has to be taken into account.

We now proceed to present the first-order corrections which, together with the results of this section, give to the formulae of sect. 2.

4. Born terms and first-order corrections

The Born terms are given by the two diagrams of fig. 1, for which the scattering amplitude reads

$$T = M_A(W) + M_B(W)$$

$$= \frac{e^2}{s}(J'_\mu J^\mu) - \frac{e^2}{M_R^2 - s}(g_V J'_\mu + g_A A'_\mu)(g_V J^\mu + g_A A^\mu)) , \qquad (4.1)$$

where

$$J_\mu = \bar{v}(q)\gamma_\mu u(p) , \qquad A_u = \bar{v}(q)\gamma_\mu\gamma_5 u(p) ,$$
$$J'_\mu = \bar{u}(p')\gamma_\mu v(q') , \qquad A'_\mu = \bar{u}(p')\gamma_\mu\gamma_5 v(q') , \qquad (4.2)$$

and $M_R^2 = M^2 - iM\Gamma$. In the standard model [2] one has (see sect. 2) $g_V^e = g_V^\mu = (4\sin^2\theta_W - 1)/2\sin(2\theta_W)$ and $g_A^e = g_A^\mu = 1/2\sin(2\theta_W)$.

The differential cross section is given by

$$\frac{d\sigma}{d\Omega} = \frac{1}{64\pi^2 s}\frac{1}{4}\sum_{\text{spin}}(TT^+) . \qquad (4.3)$$

Since we consider only transverse polarization for the $e^\pm$ and we neglect all lepton masses, in doing the trace it suffices to insert

$$\sum_{\text{spin}} u(p)\bar{u}(p) = \not{p}(1 + \not{s}_1\gamma_5) ,$$

$$\sum_{\text{spin}} v(q)\bar{v}(q) = \not{q}(1 + \not{s}_2\gamma_5) ,$$

where

$$s_1^\mu = (0, \xi_-, 0, 0), \qquad s_2^\mu = (0, \xi_+, 0, 0) , \qquad (4.4)$$

with $\xi_+\xi_- < 0$.

In the appendix, we list all the traces and two useful identities for the calculations to follow. We have written these down explicitly to facilitate a check of our resulting formulae by the reader.

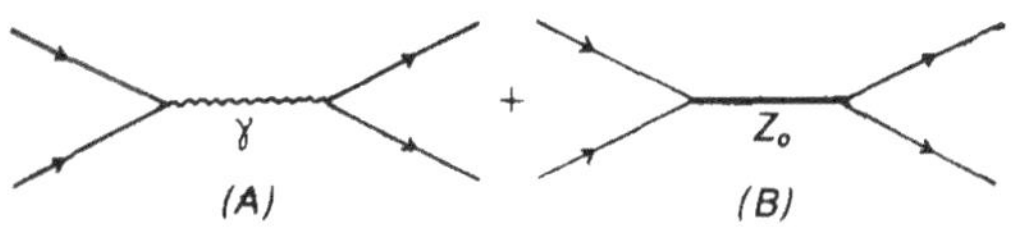

Fig. 1. Born diagrams with γ and Z_0 in the s-channel.

Using eqs. (4.1) and (4.3) and upon performing the relevant traces we find that the Born cross section may be written in the following way:

$$\left(\frac{d\sigma}{d\Omega}\right)_{\text{Born}} = \frac{\alpha^2}{4s}\{[(1+\cos^2\theta)A(s) + 2B(s)\cos\theta] - |\xi_- \xi_+|\sin^2\theta[C(s)\cos 2\phi - D(s)\sin 2\phi]\}, \quad (4.5)$$

where

$$A(s) = 1 + 2sg_V^2 \,\text{Re}\left(\frac{1}{s-M_R^2}\right) + \frac{s^2}{|s-M_R^2|^2}(g_V^2+g_A^2)^2, \quad (4.6a)$$

$$B(s) = 2sg_A^2 \,\text{Re}\left(\frac{1}{s-M_R^2}\right) + \frac{4s^2}{|s-M_R^2|^2}g_V^2 g_A^2, \quad (4.6b)$$

$$C(s) = 1 + 2sg_V^2 \,\text{Re}\left(\frac{1}{s-M_R^2}\right) + \frac{s^2}{|s-M_R^2|^2}(g_V^4-g_A^4), \quad (4.6c)$$

$$D(s) = 2sg_V g_A \,\text{Im}\left(\frac{1}{s-M_R^2}\right), \quad (4.6d)$$

which coincides with eqs. (2.9), once eqs. (2.7) are taken into account.

We now consider the first-order virtual corrections given by the interference of the diagrams in fig. 1 with the box diagrams of figs. 2 and 3.

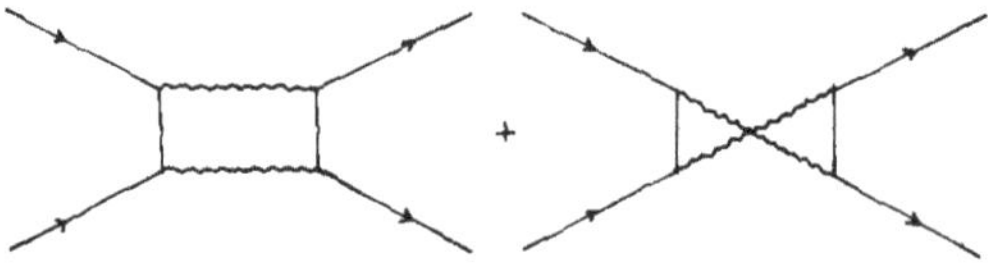

Fig. 2. 2 photon diagrams.

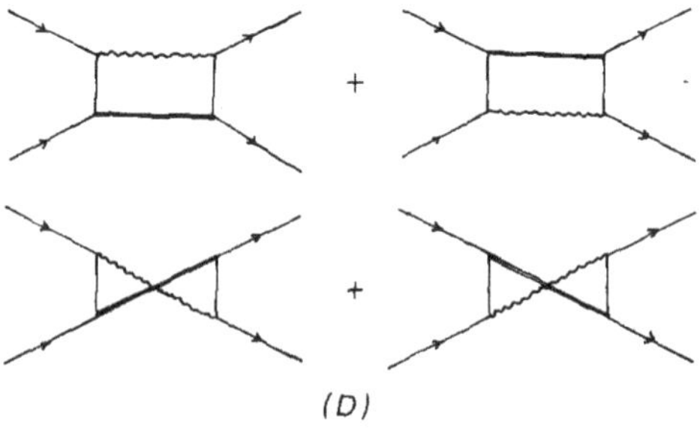

Fig. 3. $(\gamma + Z_0)$ intermediate states.

The 2-photon exchange diagrams have been calculated earlier many times [11–13]. We can thus omit the details and simply record the result:

$$M_{\mathrm{C}}=\frac{2\alpha^2}{s}\{(J'_\mu J^\mu)(V_1+2\pi iV_2)+(A'_\mu A^\mu)(A_1+2\pi iA_2)\}, \tag{4.7}$$

where

$$V_1=-8\ln\frac{a}{b}\ln\frac{2E}{\lambda}-z\left(\frac{\ln^2 a}{b^4}+\frac{\ln^2 b}{a^4}\right)+\frac{\ln a}{b^2}-\frac{\ln b}{a^2}, \tag{4.8a}$$

$$V_2=2\ln\frac{a}{b}-\tfrac{1}{2}z\left(\frac{\ln a}{b^4}+\frac{\ln b}{a^4}\right)-\frac{z}{1-z^2}, \tag{4.8b}$$

$$A_1=-z\left(\frac{\ln^2 a}{b^4}-\frac{\ln^2 b}{a^4}\right)+\frac{\ln a}{b^2}+\frac{\ln b}{a^2}, \tag{4.8c}$$

$$A_2=\tfrac{1}{2}z\left(\frac{\ln a}{b^4}-\frac{\ln b}{a^4}\right)+\frac{1}{1-z^2}, \tag{4.8d}$$

and λ denotes the mass of the photon.

The diagrams in fig. 3 have been previously computed for a resonance with vector coupling only [12, 13], e.g., J/ψ. We need to include axial-vector coupling as well. For example, the first diagram gives

$$M_{\mathrm{D}_1}=-ie^4\int\frac{\mathrm{d}^4k}{(2\pi)^4}\,\frac{[\bar{u}(p')\gamma_\mu(\not{p}'+\not{k}+m_\mu)\gamma_\mu(g_{\mathrm{V}}+g_{\mathrm{A}}\gamma_5)v(q')]\times[\bar{v}(q)\gamma^\nu(g_{\mathrm{V}}+g_{\mathrm{A}}\gamma_5)(\not{p}+\not{k}+m_e)\gamma^\mu u(p)]}{(k^2-\lambda^2)[(k+p+q)^2-M_{\mathrm{R}}^2][(p+k)^2-m_e^2][(p'+k)^2-m_\mu^2]}. \tag{4.9}$$

As before, we neglect all masses in the numerator. Also, we make the standard approximation of setting $k\to 0$ in the numerator. The terms proportional to k_μ, which will be of order $(M_{\mathrm{R}}^2/s)\ln(M_{\mathrm{R}}^2/s)$, can in fact be neglected safely in the vicinity of the resonance. This simplifies the calculation tremendously. Using the identities (A.1) and (A.2) of the appendix, we find in this limit:

$$M_{\mathrm{D}_1}=-ie^4(4p\cdot p')(g_{\mathrm{V}}J'_\mu+g_{\mathrm{A}}A'_\mu)(g_{\mathrm{V}}J^\mu+g_{\mathrm{A}}A^\mu)I_1, \tag{4.10}$$

where

$$I_1=\int\frac{d^4k}{(2\pi)^4}\frac{1}{(k^2-\lambda^2)\left[(k+p+q)^2-M_R^2\right]\left[(p+k)^2-m_e^2\right]\left[(p'+k)^2-m_\mu^2\right]}. \tag{4.11}$$

Adding up all four diagrams of fig. 3 we finally obtain

$$M_D=M_B\frac{2\alpha}{\pi}\left\{(\ln^2a-\ln^2b)+\tfrac{1}{2}\left[\mathrm{Li}_2(b^2)-\mathrm{Li}_2(a^2)\right]+\ln\frac{a}{b}\ln\frac{(M_R^2-s)^2}{\lambda^2 s}\right\}, \tag{4.12}$$

where the dilogarithm is defined as $\mathrm{Li}_2(z)=-\int_0^z dx\ln(1-x)/x$.

Now we turn to the calculation of the real emission (bremsstrahlung) diagrams. These may be separated (see fig. 4) into 2 classes, according as to whether the emission is from the initial legs, type B_i or from the final legs, type B_f. In the following, we shall indicate the product of two sets of diagrams with the symbol $B_i\otimes B_i$ or $B_i\otimes B_f$, etc. When evaluating the cross section, the products $B_i\otimes B_i$ and $B_f\otimes B_f$ in the soft photon approximation, do not depend upon the details of the interactions (i.e., whether γ or Z_0 is exchanged) and thus we may simply take over the earlier results for these [14, 15]. These terms generate corrections factors which are independent of the scattering angles (θ, ϕ) and are included in the first four terms of $C_F^{(i)}$ [eqs. (2.11)], discussed at the end of this section. On the other hand the angle-dependent parts are generated by the interference terms $B_i\otimes B_f$, a discussion of which follows.

Following ref. [13], we shall give separately the results for the sets $A\otimes C$, $B\otimes C$ and $(A+B)\otimes D$, where A, B, C and D all refer to the diagrams of fig. 4.

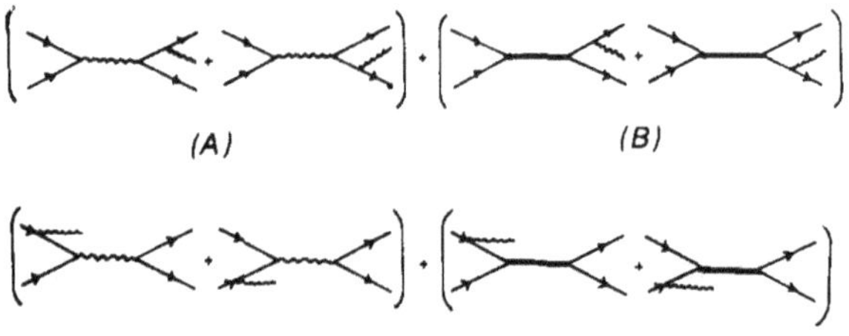

Fig. 4. Bremsstrahlung diagrams.

(i) $A \otimes C$. This term contributes to the purely QED cross section and one obtains, in the soft photon approximation,

$$\left(\frac{d\sigma}{d\Omega}\right)^{\text{real}}_{A\otimes C} = \frac{1}{64\pi^2 s} 2e^2 \int_{\omega<\Delta\omega} \frac{d^3k}{2\omega(2\pi)^3}\left(\frac{p_\mu}{p\cdot k} - \frac{q_\mu}{q\cdot k}\right)$$

$$\times\left(\frac{p'^\mu}{p'\cdot k} - \frac{q'^\mu}{q'\cdot k}\right)\tfrac{1}{4}\mathrm{Tr}\left[M_A^+(W)M_A(W)\right],$$

where $M_A(W)$ refers to fig. 1 (eq. 4.1). Then one gets

$$\left(\frac{d\sigma}{d\Omega}\right)^{\text{real}}_{A\otimes C} = \left(\frac{d\sigma_{\text{QED}}}{d\Omega}\right)\frac{4\alpha}{\pi}\left[F(z) - F(-z)\right], \tag{4.13}$$

where

$$F(z) = 4\pi^2 \int_{\omega\leqslant\Delta\omega} \frac{d^3k}{2\omega(2\pi)^3}\frac{(p\cdot p')}{(p\cdot k)(p'\cdot k)}$$

$$\simeq 2\ln\frac{2\Delta\omega}{\lambda}\ln a + \ln^2 a + \tfrac{1}{2}\mathrm{Li}_2(b^2). \tag{4.14}$$

Clearly $(z \leftrightarrow -z)$ implies $(a \leftrightarrow b)$.

To this we must add the elastic part of $A \otimes C$ (figs. 1,2) which, through eqs. (4.1), (4.7) and (4.8) is obtained to be

$$\left(\frac{d\sigma}{d\Omega}\right)^{\text{elastic}}_{A\otimes C} = \frac{1}{64\pi^2 s} 2\frac{e^2}{s}\frac{2\alpha^2}{s}$$

$$\times\tfrac{1}{4}\mathrm{Re}\left\{\mathrm{Tr}(\overline{J'\cdot J})\left[(V_1 + 2\pi i V_2)(J'\cdot J) + (A'\cdot A)(A_1 + 2\pi i A_2)\right]\right\}. \tag{4.15}$$

Summing eqs. (4.13) and (4.15) and performing the traces, we find

$$\left(\frac{d\sigma}{d\Omega}\right)_{A\otimes C} = \frac{\alpha^2}{4s}\left[1 + z^2 - P(\theta,\phi)\right]\left[1 - \frac{8\alpha}{\pi}\ln\frac{a}{b}\ln\frac{E}{\Delta\omega} + X_V(\theta,\phi)\right], \tag{4.16}$$

where [16]

$$X_V(\theta,\phi) = -\frac{\alpha}{\pi}\Bigg\{4(\ln^2 b - \ln^2 a) + 2\left[\mathrm{Li}_2(a^2) - \mathrm{Li}(b^2)\right] + z\left(\frac{\ln^2 a}{b^4} + \frac{\ln^2 b}{a^2}\right) - \left(\frac{\ln a}{b^2} - \frac{\ln b}{a^2}\right) + \frac{2z}{1+z^2-P(\theta,\phi)}\left[z\left(\frac{\ln^2 a}{b^4} - \frac{\ln^2 b}{a^4}\right) - \left(\frac{\ln a}{b^2} + \frac{\ln b}{a^2}\right)\right]\Bigg\}. \quad (4.17)$$

In eq. (4.16) the infrared factor dependent on $\ln(E/\Delta\omega)$, is the first-order expansion of $(\Delta\omega/E)^{2\beta_{\rm int}}$ [see eq. (2.10a)]. Eq. (4.17) gives the finite factor of eq. (2.12a).

(ii) $B \otimes C$. For the real part, we obtain (fig. 4)

$$\left(\frac{d\sigma}{d\Omega}\right)^{\rm real}_{B\otimes C} = \frac{1}{64\pi^2 s}\,\mathrm{Re}\Bigg\{\tfrac{1}{4}\mathrm{Tr}\big(M_A^+(W)M_B(W)\big) \times \frac{2e^2}{(2\pi)^3}\int_{\omega<\Delta\omega}\frac{d^3k}{2\omega}\left(\frac{p_\mu}{p\cdot k} - \frac{q_\mu}{q\cdot k}\right)\left(\frac{p'_\mu}{p'\cdot k} - \frac{q'_\mu}{q'\cdot k}\right)\Bigg\}, \quad (4.18)$$

which, in analogy to eq. (4.13), gives

$$\left(\frac{d\sigma}{d\Omega}\right)^{\rm real}_{B\otimes C} = \left\{\frac{d\sigma_{\rm int,V}}{d\Omega} + \frac{d\sigma_{\rm int,A}}{d\Omega} + \frac{d\sigma_{\rm int,VA}}{d\Omega}\right\} \times \frac{2\alpha}{\pi}\left[F(z) - F(-z)\right]. \quad (4.19)$$

The corresponding elastic part is obtained through eqs. (4.1), (4.7) and (4.8). We find

$$\left(\frac{d\sigma}{d\Omega}\right)^{\rm elastic}_{B\otimes C} = \frac{1}{64\pi^2 s}2\,\mathrm{Re}\left\{\tfrac{1}{4}\mathrm{Tr}\left[(M_B^+(W)M_C(W)\right]\right\}$$
$$= \frac{\alpha^3}{4\pi s}\left[V_1\,\mathrm{Re}\,\chi + 2\pi V_2\,\mathrm{Im}\,\chi\right]\left[r_V(1+z^2-P(\theta,\phi)) + 2r_A z\right] + \left[A_1\,\mathrm{Re}\,\chi + 2\pi A_2\,\mathrm{Im}\,\chi\right]\left[r_A(1+z^2+P(\theta,\phi) + 2r_V z\right] - \left[2\pi(V_2-A_2)\,\mathrm{Re}\,\chi - (V_1-A_1)\,\mathrm{Im}\,\chi\right]\left[r_{AV}Q(\theta,\phi)\right], \quad (4.20)$$

where $\chi(s)$ is defined in eq. (2.7b).

(iii) (A + B)⊗ D. For this case there is an almost complete cancellation between the real and virtual diagrams, as in the pure vector case [7, 12, 13]. For the real ones, we have

$$\left(\frac{\mathrm{d}\sigma}{\mathrm{d}\Omega}\right)^{\mathrm{real}}_{(\mathrm{A+B})\otimes\mathrm{D}} = \frac{1}{64\pi^2 s} 2e^2 \,\mathrm{Re}\left\{ \int \frac{\mathrm{d}^3 k}{2\omega(2\pi)^3} \left(\frac{p_\mu}{p\cdot k} - \frac{q_\mu}{q\cdot k} \right) \right.$$

$$\left. \times \left(\frac{p'_\mu}{p'\cdot k} - \frac{q'_\mu}{q'\cdot k} \right) \tfrac{1}{4}\left[(M_\mathrm{A}(W) + M_\mathrm{B}(W) \right]^+ M_\mathrm{B}(W-k) \right\}. \tag{4.21}$$

Using eqs. (4.1) and (4.21) we find

$$\left(\frac{\mathrm{d}\sigma}{\mathrm{d}\Omega}\right)^{\mathrm{real}}_{(\mathrm{A+B})\otimes\mathrm{D}} = \frac{1}{64\pi^2 s}\frac{\alpha}{\pi} \mathrm{Re}\left\{ \mathrm{Tr}\left[(M_\mathrm{A}(W) + M_\mathrm{B}(W))^+ M_\mathrm{B}(W) \right] \right.$$

$$\left. \times \left[F'(z) - F'(-z) \right] \right\}, \tag{4.22}$$

where

$$F'(z) = 4\pi^2 \int_{\omega < \Delta\omega} \frac{\mathrm{d}^3 k}{2\omega(2\pi)^3} \frac{(p\cdot p')}{(p\cdot k)(p'\cdot k)} \left\{ \frac{1}{1 + 4E\omega/(M_\mathrm{R}^2 - s)} \right\}$$

$$= 2\ln a \ln \frac{2\Delta\omega}{\lambda\left[1 + 4E\Delta\omega/(M_\mathrm{R}^2 - s) \right]} + \ln^2 a + \tfrac{1}{2}\,\mathrm{Li}_2(b^2). \tag{4.23}$$

For the corresponding elastic part (figs. 1–3)

$$\left(\frac{\mathrm{d}\sigma}{\mathrm{d}\Omega}\right)^{\mathrm{elastic}}_{(\mathrm{A+B})\otimes\mathrm{D}} = \frac{2}{64\pi^2 s}\frac{1}{4} \mathrm{Re}\left\{ \mathrm{Tr}\left[(M_\mathrm{A}(W) + M_\mathrm{B}(W))^+ M_\mathrm{D}(W) \right] \right\},$$

we obtain, through eqs. (4.1) and (4.12),

$$\left(\frac{\mathrm{d}\sigma}{\mathrm{d}\Omega}\right)^{\mathrm{elastic}}_{(\mathrm{A+B})\otimes\mathrm{D}} = -\frac{1}{64\pi^2 s}\frac{\alpha}{\pi} \mathrm{Re}\left\{ \mathrm{Tr}\left[(M_\mathrm{A}(W) + M_\mathrm{B}(W))^+ M_\mathrm{B}(W) \right] \right.$$

$$\left. \times \left[2\ln\frac{a}{b} \ln \frac{M_\mathrm{R}^2 - s}{2\lambda E} + \tfrac{1}{2}\left(\mathrm{Li}_2(b^2) - \mathrm{Li}_2(a^2) \right) + \ln^2 a - \ln^2 b \right] \right\}. \tag{4.24}$$

Summing the real and virtual contributions given by eqs. (4.22) and (4.24) we get the final result

$$\left(\frac{d\sigma}{d\Omega}\right)_{(A+B)\otimes D}=\frac{1}{64\pi^2 s}\frac{\alpha}{\pi}\operatorname{Re}\left\{\operatorname{Tr}\left[(M_A(W)+M_B(W))^+M_B(W)\right]\right.$$

$$\left.\times\left[2\ln\frac{a}{b}\ln\left(1+\frac{M_R^2-s}{4E\Delta\omega}\right)\right]\right\}, \tag{4.25}$$

which is negligible when $\Delta\omega\gg|(M_R^2-s)|/4E$.

The total angle-dependent first-order correction factor is given by the sum of eqs. (4.16), (4.19), (4.20) and (4.25). It is easily verified that it checks with eqs. (2.10a, b, c), when the infrared factors are expanded to first order in $\beta_{\rm int}$, and with the expressions for X_i ($i=$ A, VA) and Y_j ($j=$ V, A) given in eqs. (2.12a–e) upon factoring out the corresponding Born terms. The only exception is given by the $\overline{R}$ terms, which arise from the imaginary part of the propagator correction and are discussed below in connection with the vacuum polarization factor.

The first-order correction $\Pi(s,\, m_i^2)$ to the photon propagator $\Pi(s)$, due to a fermion loop of charge eQ_i and mass m_i, is given in the relativistic limit ($s\gg m_i^2$) by

$$\Pi(s)=1+\Pi(s,\, m_i^2), \tag{4.26}$$

with

$$\operatorname{Re}\Pi(s,m_i^2)\simeq\frac{\alpha}{3\pi}\left[\ln\frac{s}{m_i^2}-\frac{5}{3}\right] \tag{4.27}$$

and

$$\operatorname{Im}\,\Pi(s,\, m_i^2)\simeq-\tfrac{1}{3}\alpha Q_i^2\,. \tag{4.28}$$

Thus, the vacuum polarization correction due to the ith fermion loop is

$$\left(\frac{d\sigma}{d\Omega}\right)_{\rm VP}^{(i)}=\frac{1}{64\pi^2 s}2\operatorname{Re}\left\{\left[\operatorname{Re}\Pi(s,m_i^2)-i\operatorname{Im}\Pi(s,m_i^2)\right]\right.$$

$$\left.\times\tfrac{1}{4}\operatorname{Tr}\left[M_A^+(W)(M_A(W)+M_B(W))\right]\right\}, \tag{4.29}$$

which leads to

$$\left(\frac{d\sigma}{d\Omega}\right)^{(i)}_{VP}=\frac{\alpha^2}{4s}\{2\,\mathrm{Re}\,\Pi(s,m_i^2)\{(1+z^2-P(\theta,\phi))+\mathrm{Re}\,\chi$$

$$\times\left[r_V(1+z^2-P(\theta,\phi))+2zr_A\right]+\mathrm{Im}\,\chi\cdot r_{AV}Q(\theta,\phi)\}$$

$$+2\,\mathrm{Im}\,\Pi(s,m_i^2)\{\mathrm{Im}\,\chi[r_V(1+z^2-P(\theta,\phi))+2zr_A]$$

$$-\mathrm{Re}\,\chi\cdot r_{AV}Q(\theta,\phi)\}\}\,. \tag{4.30}$$

In the total Re $\Pi(s)$ we shall include the contributions of the leptons e, μ, τ as well as the hadronic part, which has been calculated in refs. [17, 18]. In particular explicit use of the experimental data has been made in the dispersive integral for $\Pi_{had}(s)$ at low energies ($\sqrt{s'}\lesssim 9$ GeV), while the high-energy part has been estimated through the six quarks u, d, s, c, b and t. This leads to a total correction factor

$$\delta_{VP}^{tot}=2\mathrm{Re}[\Pi(s)-1]=2\sum_i \mathrm{Re}\Pi(s,m_i^2)$$

$$=\delta_{VP}(m_e^2)+\delta_{VP}(m_\mu^2)+\delta_{VP}, \tag{4.31}$$

where

$$\delta_{VP}=(1.86\ln W-1)\cdot 10^{-2}-\frac{2\alpha}{3\pi}\left[\tfrac{1}{3}I_b(W)+\tfrac{4}{3}I_t(W)\right],$$

$$I_i(W)=\frac{5}{3}+\frac{4m_i^2}{W^2}-\sqrt{1-\frac{4m_i^2}{W^2}}\left(1+\frac{2m_i^2}{W^2}\right)\ln\left\{\frac{W}{2m_i}\left[1+\sqrt{1-\frac{4m_i^2}{W^2}}\right]\right\}, \tag{4.32}$$

and we have separated out the electron and muon terms.

On the other hand, the total contribution to Im $\Pi(s)$ is simply obtained as $-\frac{1}{3}\alpha\bar{R}$ where $\bar{R}=\Sigma_{i=\mathrm{leptons+quarks}}\,Q_i^2$. With all that in mind, eqs. (4.30) lead to the vacuum polarization factors included in eqs. (2.11) and (2.12).

Finally, finite vertex corrections and bremsstrahlung terms independent of the scattering angle are contained in the first four terms of $C_F^{(i)}$ [see eqs. (2.11)], which also include, as discussed above, the electron and muon vacuum polarization factors $\delta_{VP}(m_e^2)$ and $\delta_{VP}(m_\mu^2)$. The former terms are well known and can be found, for instance, in refs. [7, 15].

This completes our discussion of finite first-order corrections.

5. Discussion and numerical results

We have presented a detailed study of higher-order electromagnetic effects in the reaction $e^+e^- \to Z_0 \to \mu^+\mu^-$, in the framework of the Weinberg-Salam model, for the case of unpolarized as well as transversely polarized electron positron beams. Infrared factors have been considered to all orders in α, together with full finite first-order corrections. Only hard bremsstrahlung effects, other than the tail effect, have not been taken into account. The weak effects are considered to lowest order, the weak boson Z_0 being taken as a resonance of mass M and width Γ. Our results are then given by simple analytical expressions in terms of s, M, Γ and $\Delta\omega$, of immediate experimental application. This formalism, which certainly describes with very good accuracy the regions below, around and not too far above the Z_0, can be extended at larger energies by adding first-order pure weak corrections [4].

In order to explicitly show the importance of the radiative effects we have calculated, we present in figs. 5–7 some numerical results for $(d\sigma/d\Omega)$, σ and the integrated forward-backward asymmetry, in case of unpolarized beams and for the following choice of the parameters.

We have considered 6 quarks and 6 leptons and assumed for simplicity $\sin^2\theta_W = \frac{1}{4}$. This gives $M \simeq 86.4$ GeV, $\Gamma \simeq 2.2$ GeV and $\Gamma(Z_0 \to e\bar{e}) \simeq 70$ MeV. Furthermore we have used for $\Delta = \Delta\omega/E$ the values 5×10^{-3}, 10^{-2}, 5×10^{-2} and 10^{-1}. The comparison is always made with the Born terms called "naive" in the figures.

From inspection of figs. 5–7 it is clear that e.m. radiative corrections substantially change the naive results. The effect on the asymmetry, in particular, is rather

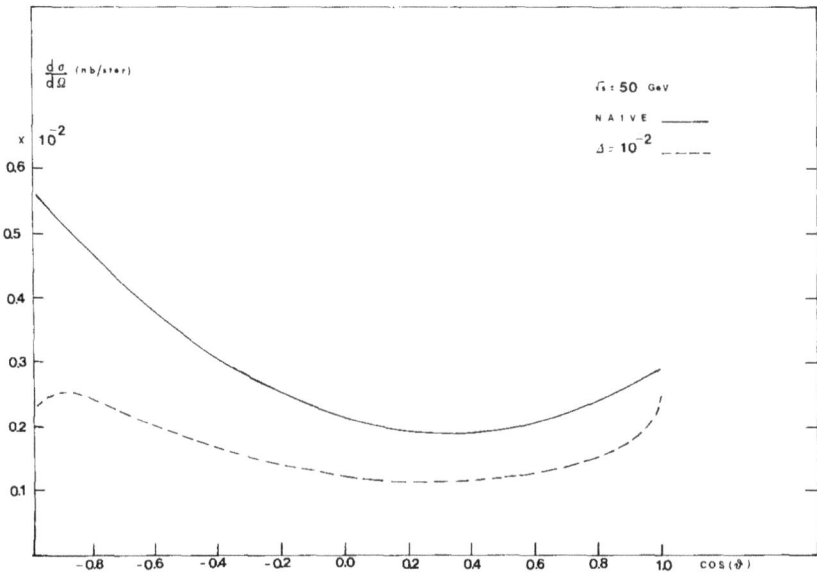

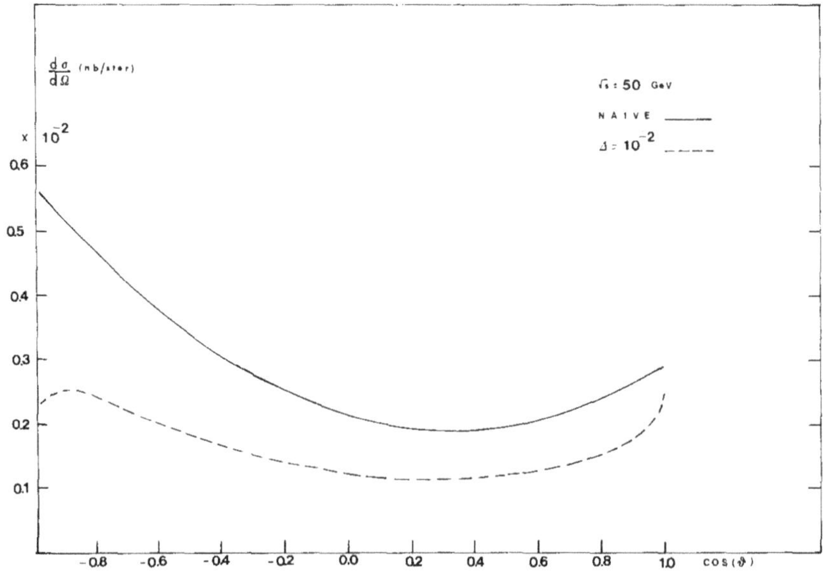

Fig. 5a.

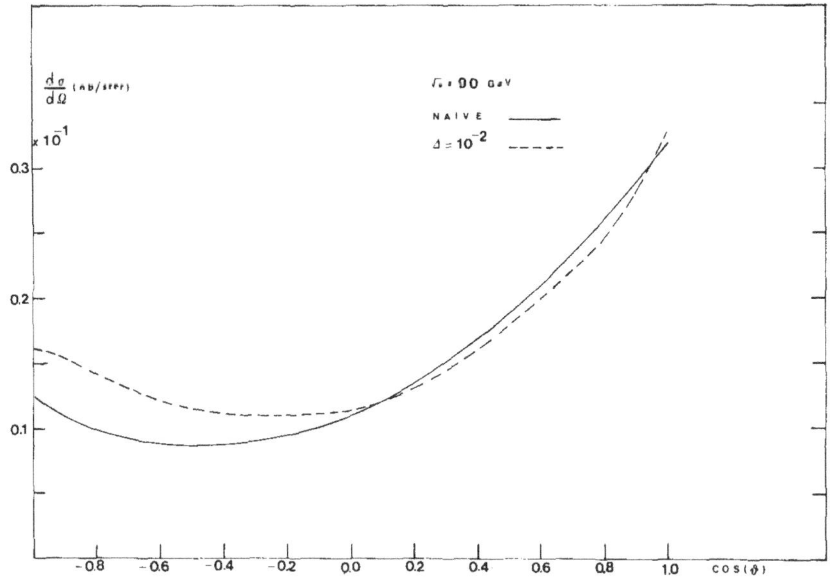

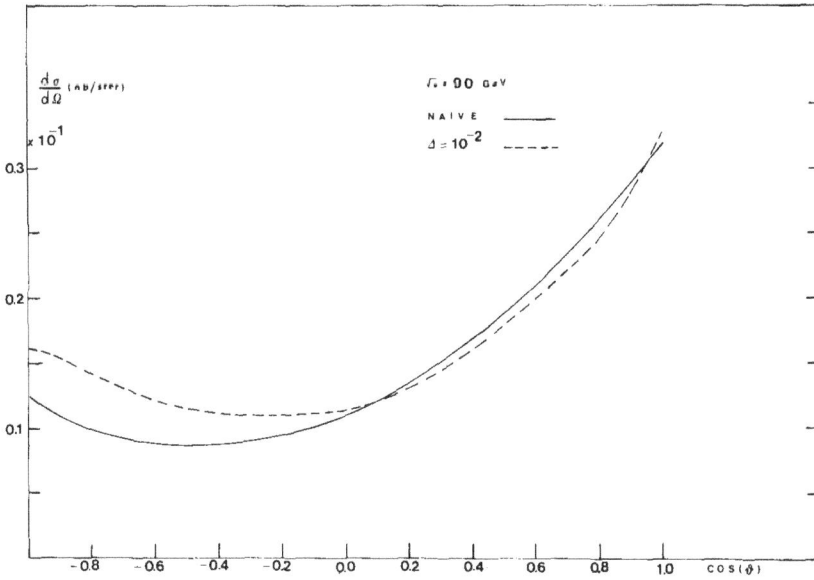

(b)

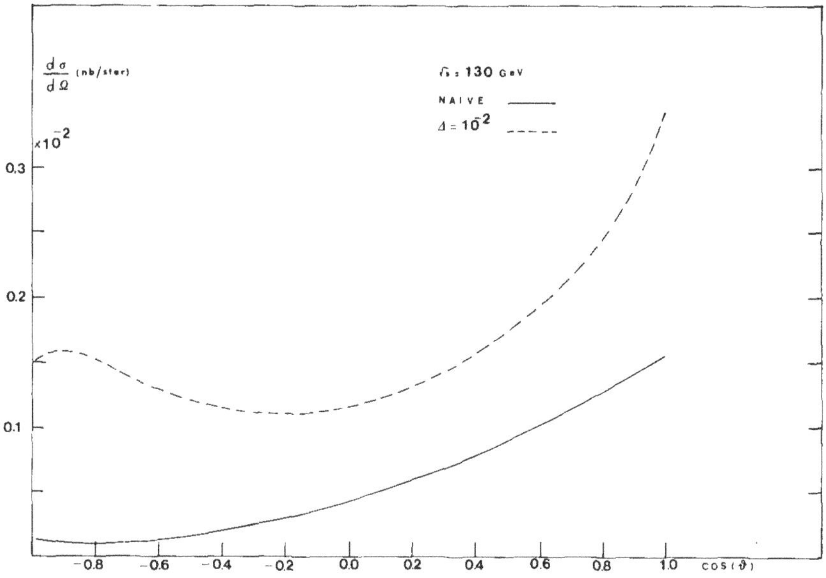

(c)

Fig. 5. Differential cross section $d\sigma/d\Omega$ versus $\cos\theta$, for various energies, with and without radiative corrections (naive).

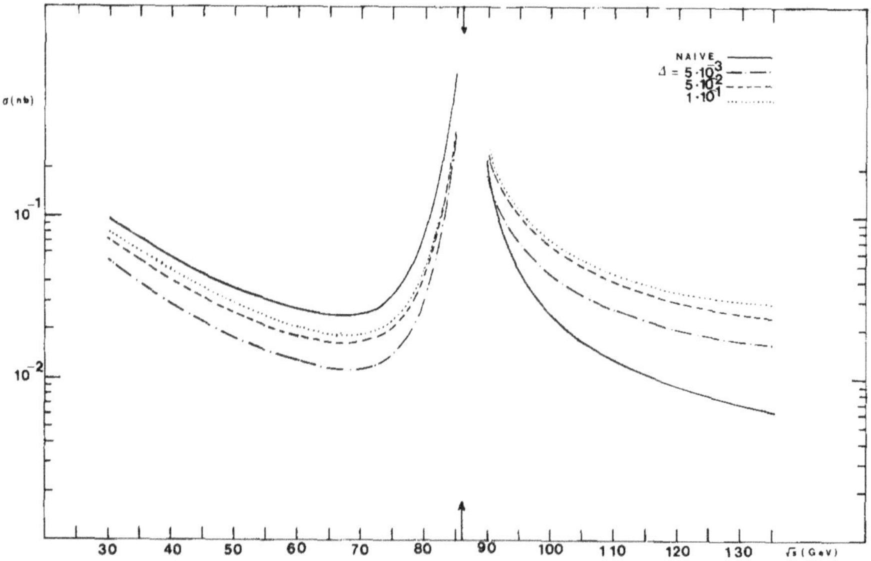

Fig. 6. Total cross section σ versus $\sqrt{s}$ with and without radiative corrections (naive). Various values of the experimental resolution $\Delta\omega/E = \Delta$ are considered.

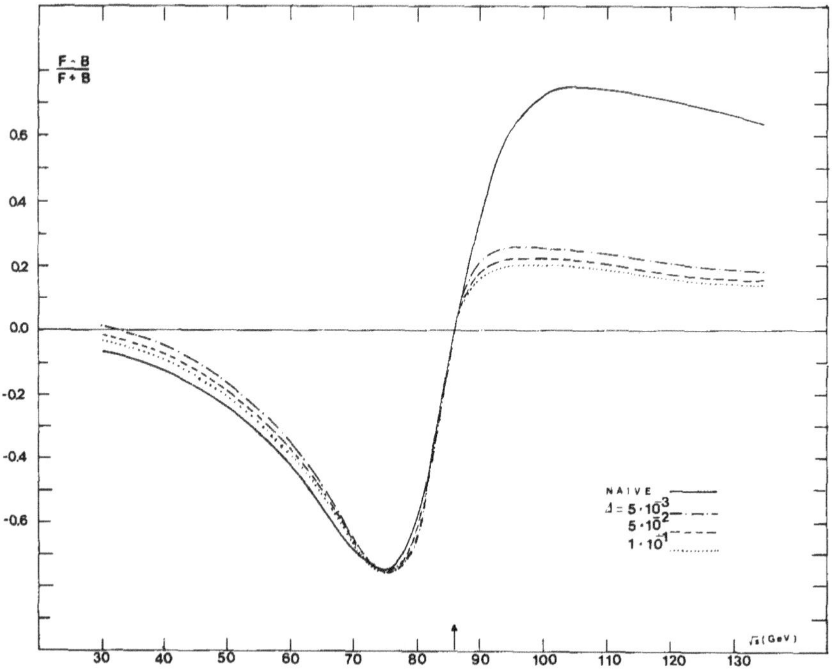

Fig. 7. Integrated asymmetry versus $\sqrt{s}$, for various values of the experimental resolution (dotted lines) and without radiative corrections (full line).

impressive for energies above the Z_0 mass. It follows, therefore, that one has to take full account of radiative effects even for planning experiments at LEP energies.

Similar consideration of course apply to all reactions proceeding from e^+e^- annihilation in the vicinity of the Z_0, including purely hadronic final states.

Appendix

Here we record two identities and the trace formulae which are needed to obtain our final expressions.

For *massless* spinors the identity

$$(p'\cdot J)(p\cdot J') + (p\cdot A')(p'\cdot A) = (p\cdot p')[(J'\cdot J) + (A'\cdot A)] \qquad \text{(A.1)}$$

can be found in refs. [12, 13]. For our purposes, we need another identity involving $(J'\cdot A)$-type terms. We have obtained the following, which we quote here without proof:

$$(p\cdot J')(p'\cdot A) + (p\cdot A')(p'\cdot J) = (p\cdot p')[(J'\cdot A) + (A'\cdot J)]\,. \qquad \text{(A.2)}$$

Eqs. (A.1) and (A.2) are useful in reducing the bow diagram amplitudes of eq. (4.9) to the form given in eq. (4.10).

In the following we write down all the expressions we used for computing the traces:

$$\mathrm{Tr}\{(\overline{J'.J})(J'.J)\} = \mathrm{Tr}\{(\overline{A'.J})(A'.J)\}$$

$$= 4s^2(1 + z^2 - |\xi_+\xi_-|\sin^2\theta\cos 2\phi)\,, \qquad \text{(A.3)}$$

$$\mathrm{Tr}\{(\overline{A'.A})(A'.A)\} = \mathrm{Tr}\{(\overline{J'.A})(J'.A)\} = 4s^2(1 + z^2 + |\xi_+\xi_-|\sin^2\theta\cos 2\phi)\,, \qquad \text{(A.4)}$$

$$\mathrm{Tr}\{(\overline{J'.A})(J'.J)\} = \mathrm{Tr}\{(\overline{A'.A})(A'.J)\} = +(4is^2)|\xi_+\xi_-|\sin^2\theta\sin 2\phi\,, \qquad \text{(A.5)}$$

$$\mathrm{Tr}\{(\overline{J'.J})(A'.A)\} = \mathrm{Tr}\{(\overline{J'.A})(A'.J)\} = 8s^2z\,, \qquad \text{(A.6)}$$

$$\mathrm{Tr}\{(\overline{A'.J})(J'.J)\} = \mathrm{Tr}\{(\overline{J'.A})(A'.A)\} = 0\,. \qquad \text{(A.7)}$$

References

[1] PEP Summer Study, Stanford 1974 and 1975;
Discussion Meeting on PETRA Experiments, Frascati, 1976;
M. Camilleri et al., Physics with very high energy e^+e^- colliding beams, CERN 76-18, (1976)

[2] S. Weinberg, Phys. Rev. Lett. 19 (1967) 1264;
A. Salam, Proc. 8th Nobel Symp., Stockholm 1968, ed. N. Svartholm (Almqvist and Wiksell)
[3] G.'t Hooft, Nucl. Phys. B35 (1971) 167
[4] G. Passarino and M. Veltman, Nucl. Phys. B160 (1979) 151
[5] M. Consoli, Nucl. Phys. B160 (1979) 208
[6] M. Greco, G. Pancheri-Srivastava and Y. Srivastava, Phys. Lett. 56B (1975) 367;
D.R. Yennie, Phys. Rev. Lett. 34 (1975) 239
[7] M. Greco, G. Pancheri-Srivastava and Y. Srivastava, Nucl. Phys. B101 (1975) 234
[8] M. Greco, G. Pancheri-Srivastava and Y. Srivastava, ECFA-LEP-SS G/7/9, Frascati preprint LNF-79/20(R) (March, 1979); ECFA-LEP-SS G/7/19, Frascati preprint LNF-79/63(R) (September, 1979)
[9] M. Gourdin, ECFA-LEP-SS G/7/3 and PAR-LPTHE 78/17, Paris (November, 1978)
[10] V. Chung, Phys. Rev. B140 (1965) 1110;
M. Greco and G. Rossi, Nuovo Cim. 50 (1967) 168;
N. Papanicolau, Phys. Reports 24 (1976) 229
[11] I.B. Kriplovich, Sov. J. Nucl. Phys. 17 (1973) 298
[12] G. Altarelli, R. Petronzio and R.K. Ellis, Nuovo Cim. Lett. 13 (1975) 393
[13] A.B. Kraemmer and B. Lautrup, Nucl. Phys. B95 (1975) 380
[14] G. Pancheri, Nuovo Cim. 60 (1969) 321
[15] F.A. Berends, K.J.F. Gaemers and R. Gastmans, Nucl. Phys. B57 (1973) 381
[16] M. Greco and A.F. Grillo, Nuovo Cim. Lett. 15 (1976) 174
[17] F.A. Berends and G.J. Komen, Phys. Lett. 63B (1976) 432
[18] F. Yndurain, Nucl. Phys. B136 (1978) 533

Nuclear Physics B197 (1982) 543–546

ERRATA

M. Greco, G. Pancheri and Y. Srivastava, Radiative Corrections to $e^+e^- \to \mu^+\mu^-$ around the Z_0, Nucl. Phys. B171 (1980) 118.

Eq. (2.6a) should read

$$r_V = \frac{g_V^2}{g_V^2 + g_A^2} = \frac{(1-4\sin^2\theta_W)^2}{1+(1-4\sin^2\theta_W)^2}$$

with $g_V = (4\sin^2\theta_W - 1)/2\sin(2\theta_W)$ and $g_A = 1/2\sin(2\theta_W)$.

Eq. (4.8d) should read

$$A_2 = -\tfrac{1}{2}z\left(\frac{\ln a}{b^4} - \frac{\ln b}{a^4}\right) + \frac{1}{1-z^2}\,.$$

Eq. (4.12) should carry the opposite sign.

Eq. (4.32) should read

$$I_l(W) = \tfrac{5}{3} + \frac{4m_l^2}{W^2} - 2\sqrt{1-\frac{4m^2}{W^2}}\left(1+\frac{2m_l^2}{W^2}\right)\ln\left\{\frac{W}{2m_l}\left[1+\sqrt{1-\frac{4m_l^2}{W^2}}\right]\right\},$$

Due to an error in the computer program our figs. 5–7 are wrong and should be replaced by the following ones. Notice that we have chosen different values of Δ to make them more realistic.

We thank Drs. F. Berends and R. Kleiss for calling our attention to the numerical calculations.

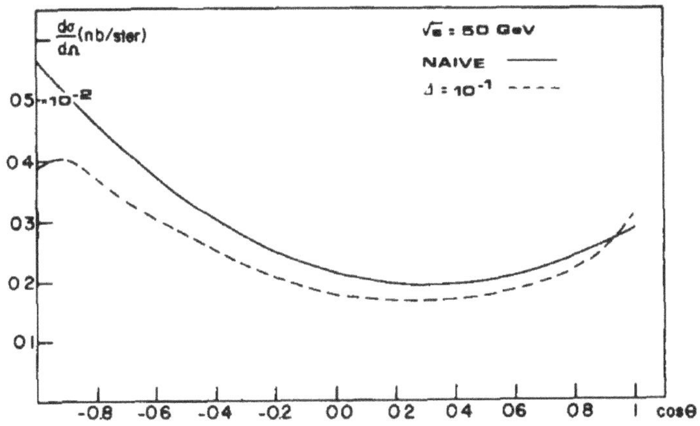

Fig 5a

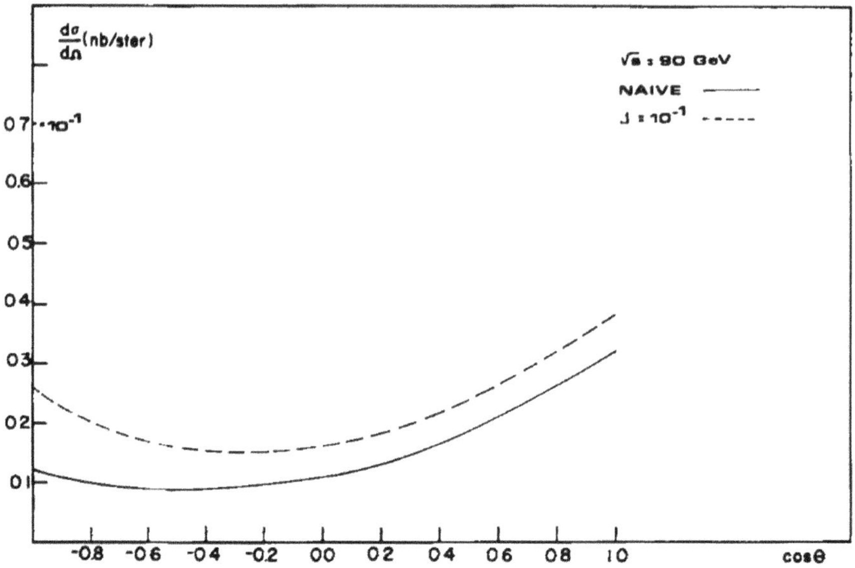

$\frac{d\sigma}{d\Omega}$(nb/ster)
$\sqrt{s}$ = 90 GeV
NAIVE
Δ = 10⁻¹
0.7 × 10⁻¹
0.6
0.5
0.4
0.3
0.2
0.1
-0.8
-0.6
-0.4
-0.2
0.0
0.2
0.4
0.6
0.8
1.0
cosθ

Fig 5b.

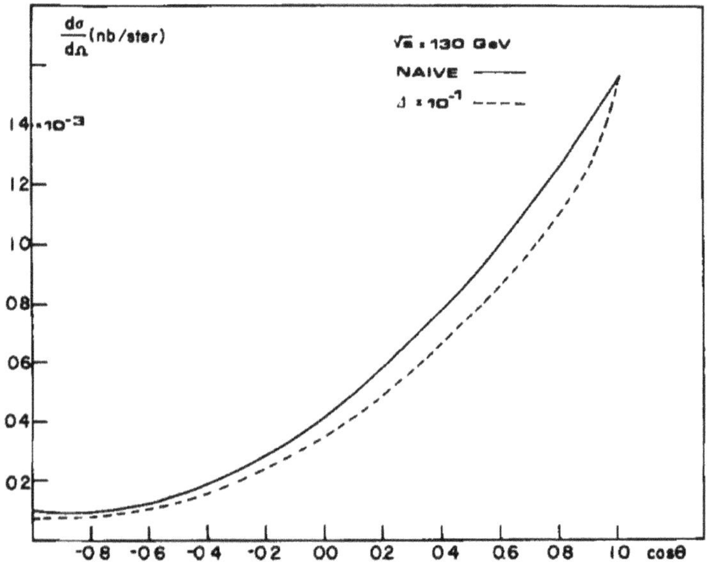

$\frac{d\sigma}{d\Omega}$(nb/ster)
$\sqrt{s}$ = 130 GeV
NAIVE
Δ = 10⁻¹
1.4 × 10⁻³
1.2
1.0
0.8
0.6
0.4
0.2
-0.8
-0.6
-0.4
-0.2
0.0
0.2
0.4
0.6
0.8
1.0
cosθ

Fig 5c

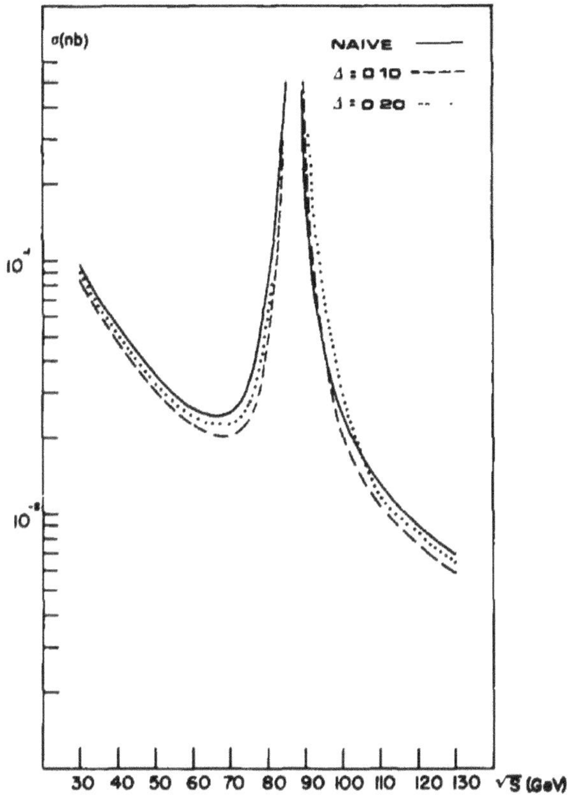
σ(nb)
NAIVE
Δ = 0.10
Δ = 0.20
30 40 50 60 70 80 90 100 110 120 130 √S (GeV)

Fig 6

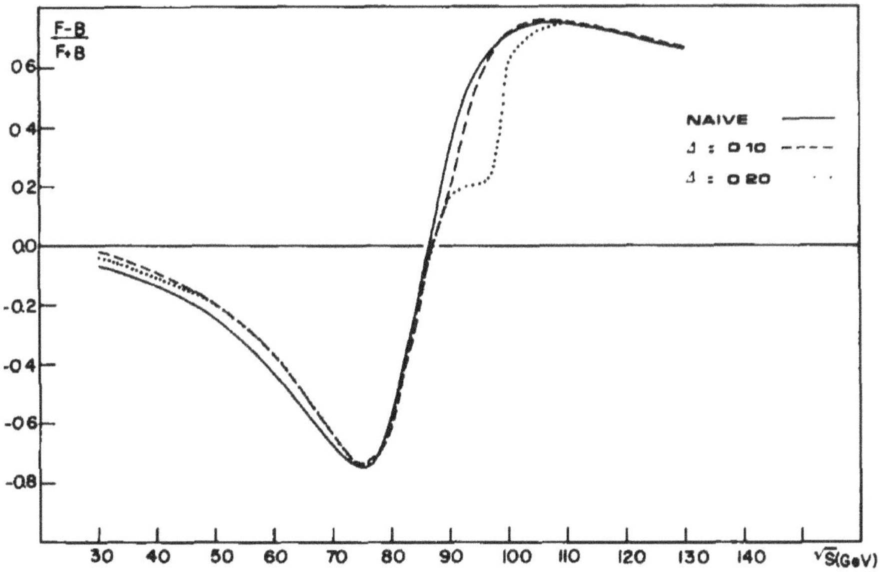
F-B / F+B
0.6
0.4
0.2
0.0
-0.2
-0.4
-0.6
-0.8
NAIVE
Δ : 0.10
Δ : 0.20
30 40 50 60 70 80 90 100 110 120 130 140 √S(GeV)

Fig 7

Volume 228, number 1 PHYSICS LETTERS B 7 September 1989

COHERENT STATES AND STRUCTURE FUNCTIONS IN QED

F. AVERSA and M. GRECO
INFN, Laboratori Nazionali di Frascati, C.P. 13, I-00044 Frascati, Italy

Received 26 May 1989

The methods of coherent states and of structure functions in QED are considered in detail for a precise evaluation of the radiative effects at LEP/SLC. They are explicitly shown to give identical results for the exponentiated infrared factors and the $o(\alpha)$ terms corresponding to the initial and final state radiation. Then the combined use of both techniques further improves the theoretical predictions to an accuracy better than 1% – including interference effects – and provides simple analytical formulae of immediate phenomenological application.

The role played by QED radiative corrections in e^+e^- reactions at LEP/SLC energies for the precise determination of electroweak parameters is well known [#1]. The resummation of the leading double and single logarithms of soft origin [3], together with the exact determination of the left-over $o(\alpha)$ terms [4,5], is essential to reach a degree of accuracy of $o(1\%)$ in the theoretical predictions. The calculation of all logarithmic $o(\alpha^2)$ corrections, as recently done [6,7] for the reaction $e^+e^-\to\mu^+\mu^-$, is the further necessary tool to get absolute control of the QED corrections, as required for precision tests of the theory.

To this aim the method of the structure functions [8], extended to both initial and final states [6] in the reaction $e^+e^-\to\mu^+\mu^-$, has been shown to be quite powerful, suggesting a systematic approach to other processes as, for example, Bhabha scattering. However, only numerical solutions have been obtained so far in the case of a resonant cross section, leaving unclear the connection to other analytical approaches to the problem. The method of coherent states [9], for example, has been proved to be very successful in describing the main features of multiphoton emission with good accuracy, giving explicit analytical results for resonant processes and interference effects with pure QED background [3,4]. The determination of the left-over $o(\alpha^2)$ logarithms is needed only for the evaluation of the radiative effects beyond the 1% level.

A close connection between the two above methods of resummation has been explicitly shown [10] to exist in the non-singlet case for QED non-resonant processes, as well as for QCD. The aim of the present paper is to further explore this relation for e^+e^- reactions when they proceed via Z exchange also. More in detail we will explicitly show that the method of structure functions applied to initial and final state radiation coincides with the coherent state approach with an accuracy of $o(1\%)$, giving explicitly in addition the desired $o(\alpha^2)$ left-over corrections.

The basic formula which describes the reaction $e^+e^-\to\mu^+\mu^-$ in the approach of structure functions is the following [6]:

$$\sigma(s)=\int_0^{\epsilon} dx\,\sigma_0((1-x)s)H_e(x,s)F_\mu(\epsilon-x,(1-x)s), \tag{1}$$

where $\epsilon=\Delta E/E$ and the initial and final state radiation kernels are expressed in terms of the electron and muon structure functions as

[#1] See ref. [1] for recent reviews of LEP/SLC physics. See ref. [2] for a review of radiative corrections at LEP/SLC.

$$H_e(x,s)=\int_{1-x}^{1}\frac{dz}{z}D_e(z,s)D_e\left(\frac{1-x}{z},s\right),\quad F_\mu(x,s)=\int_0^x dy\,H_\mu(y,(1-y)s), \tag{2,3}$$

and $H_\mu(x,s)$ is defined as in eq. (2). By taking into account the effect of the soft radiation to all orders and of the hard one up to $o(\alpha^2)$ one obtains [#2]

$$H_e(x,s)=\Delta_e(s)\beta_e x^{\beta_e-1}-\tfrac{1}{2}\beta_e(2-x)+\tfrac{1}{8}\beta_e^2\left((2-x)[3\ln(1-x)-4\ln x]-4\frac{\ln(1-x)}{x}+x-6\right), \tag{4}$$

with $\beta_e=(2\alpha/\pi)(L_e-1)$, $L_e=\ln(s/m_e^2)$ and

$$\Delta_e(s)=1+\frac{\alpha}{\pi}\{\tfrac{3}{2}L_e+2[\zeta(2)-1]\}$$
$$+\left(\frac{\alpha}{\pi}\right)^2\{[\tfrac{9}{8}-2\zeta(2)]L_e^2+[3\zeta(3)+\tfrac{11}{2}\zeta(2)-\tfrac{45}{16}]L_e+[-\tfrac{6}{5}\zeta^2(2)-\tfrac{9}{2}\zeta(3)-6\zeta(2)\ln 2+\tfrac{3}{8}\zeta(2)+\tfrac{57}{12}]\}$$
$$\equiv 1+\frac{\alpha}{\pi}\Delta_e^{(1)}+\left(\frac{\alpha}{\pi}\right)^2\Delta_e^{(2)}. \tag{5}$$

By insertion of (4) in eqs. (2) and (1) one easily obtains

$$\sigma(s)=\int_0^\epsilon dx\,\sigma_0(s(1-x))[\Delta_e(s)\Delta_\mu(s)\beta_e x^{\beta_e-1}(\epsilon-x)^{\beta_\mu}+R(x,\ldots)], \tag{6}$$

where the first term on the RHS of eq. (6) is proportional to the leading soft contribution, while $R(x,\ldots)$ gives further correction terms of order (β^2) and $\beta\epsilon$ in the final cross section. We will assume the fractional energy resolution $\epsilon\equiv\Delta E/E$ of order 10^{-1}–10^{-2}.

By splitting the Born cross section $\sigma_0(s)$ as $\sigma_0=\sigma_0^{QED}+\sigma_0^{INT}+\sigma_0^{RES}$, with [#3]

$$\sigma_0^{QED}(s)=A\frac{1}{s},\quad \sigma_0^{INT}(s)=B\,\mathrm{Re}\left(\frac{1}{s-M^2+i\Gamma M}\right),\quad \sigma_0^{RES}(s)=C\frac{s}{(s-M^2)^2+\Gamma^2M^2}, \tag{7}$$

the corresponding radiatively corrected cross sections are obtained as follows:

$$\sigma^{QED}(s)=\frac{A}{s}\left(\Delta_e(s)\Delta_\mu(s)\epsilon^{\beta_e+\beta_\mu}\frac{\Gamma(1+\beta_e)\Gamma(1+\beta_\mu)}{\Gamma(1+\beta_e+\beta_\mu)}\,{}_2F_1(1,\beta_e;1+\beta_e+\beta_\mu;\epsilon)+\ldots\right), \tag{8}$$

$$\sigma^{INT}(s)=B\left\{\Delta_e(s)\Delta_\mu(s)\epsilon^{\beta_e+\beta_\mu}\frac{\Gamma(1+\beta_e)\Gamma(1+\beta_\mu)}{\Gamma(1+\beta_e+\beta_\mu)}\right.$$
$$\left.\times\mathrm{Re}\left[\frac{1}{s-M_R^2}(1-z)^{-\beta_e}\,{}_2F_1\left(\beta_e+\beta_\mu,\beta_e;1+\beta_e+\beta_\mu;\frac{z}{z-1}\right)\right]+\ldots\right\}, \tag{9}$$

[#2] Eq. (4) agrees with the complete second order result of ref. [7] up to hard terms of $o(\alpha^2L_e)$. In the resonant region this is not of numerical importance. The generalization of our results using the more complete but longer expression of ref. [7] is straightforward.

[#3] The modifications of the Born cross sections (7) due to electroweak corrections, with the usual replacement in M and Γ, can be correspondingly introduced in our final expressions.

$$\sigma^{\mathrm{RES}}(s)=-C\left\{\frac{s}{M\Gamma}\Delta_e(s)\Delta_\mu(s)\epsilon^{\beta_e+\beta_\mu}\frac{\Gamma(1+\beta_e)\Gamma(1+\beta_\mu)}{\Gamma(1+\beta_e+\beta_\mu)}\right.$$

$$\left.\times\mathrm{Im}\left[\frac{1}{s-M_R^2}(1-z)^{-\beta_e}\,{}_2F_1\left(\beta_e+\beta_\mu,\beta_e;1+\beta_e+\beta_\mu;\frac{z}{z-1}\right)\right]+\ldots\right\},\qquad(10)$$

where $M_R^2=M^2-\mathrm{i}\Gamma M$, $z=\epsilon s/(s-M_R^2)$ and the dots indicate next to leading terms corresponding to $R(x,\ldots)$ in eq. (6). By expanding in β_e the hypergeometric functions one then finds

$$\sigma^{\mathrm{QED}}(s)=\sigma_0^{\mathrm{QED}}(s)\Delta_e(s)\Delta_\mu(s)\epsilon^{\beta_e+\beta_\mu}+\ldots,\qquad(11)$$

$$\sigma^{\mathrm{INT}}(s)=\sigma_0^{\mathrm{INT}}(s)\Delta_e(s)\Delta_\mu(s)\epsilon^{\beta_\mu}\frac{\Gamma(1+\beta_e)\Gamma(1+\beta_\mu)}{\Gamma(1+\beta_e+\beta_\mu)}\frac{1}{\cos\delta_R}\mathrm{Re}\left[\exp(\mathrm{i}\delta_R)\left(\frac{\epsilon}{1+(\epsilon s/M\Gamma)\sin\delta_R\exp(\mathrm{i}\delta_R)}\right)^{\beta_e}\right]$$

$$+\ldots,\qquad(12)$$

$$\sigma^{\mathrm{RES}}(s)=\sigma_0^{\mathrm{RES}}(s)\Delta_e(s)\Delta_\mu(s)\epsilon^{\beta_\mu}\frac{\Gamma(1+\beta_e)\Gamma(1+\beta_\mu)}{\Gamma(1+\beta_e+\beta_\mu)}\left|\frac{\epsilon}{1+(\epsilon s/M\Gamma)\sin\delta_R\exp(\mathrm{i}\delta_R)}\right|^{\beta_e}$$

$$\times(\cos\beta_e\phi-\cot\delta_R\sin\beta_e\phi)+\ldots,\qquad(13)$$

where

$$(M_R^2-s)^{-1}=\frac{\sin\delta_R\exp(\mathrm{i}\delta_R)}{M\Gamma},\quad\tan\delta_R=\frac{M\Gamma}{(M^2-s)},\quad\phi=\arctan\left(\frac{\epsilon s+M^2-s}{M\Gamma}\right)-\arctan\left(\frac{M^2-s}{M\Gamma}\right).$$

We now wish to comment briefly on the various factors appearing in the RHS of eqs. (11)–(13) and compare them with the analogous expressions obtained in the framework of coherent states, where, to leading order [#4], the differential cross section is given by [2,4]

$$\frac{\mathrm{d}\sigma}{\mathrm{d}\Omega}=\sum_i\frac{\mathrm{d}\sigma_0^{(i)}}{\mathrm{d}\Omega}\left(C_{\mathrm{infra}}^{(i)}+C_F^{(i)}\right),\quad(i=\mathrm{QED,\ RES,\ INT}),\qquad(14)$$

with

$$C_{\mathrm{infra}}^{(\mathrm{QED})}=\epsilon^{\beta_e+\beta_\mu+2\beta_{\mathrm{int}}},\qquad(15)$$

$$C_{\mathrm{infra}}^{(\mathrm{INT})}=\frac{\epsilon^{\beta_\mu+\beta_{\mathrm{int}}}}{\cos\delta_R}\mathrm{Re}\left[\exp(\mathrm{i}\delta_R)\left(\frac{\epsilon}{1+(\epsilon s/M\Gamma)\sin\delta_R\exp(\mathrm{i}\delta_R)}\right)^{\beta_e}\left(\frac{\epsilon}{\epsilon+(M\Gamma/s)\exp(-\mathrm{i}\delta_R)/\sin\delta_R}\right)^{\beta_{\mathrm{int}}}\right],\qquad(16)$$

$$C_{\mathrm{infra}}^{(\mathrm{RES})}=\epsilon^{\beta_\mu}\left|\frac{\epsilon}{1+(\epsilon s/M\Gamma)\sin\delta_R\exp(\mathrm{i}\delta_R)}\right|^{\beta_e}\left|\frac{\epsilon}{\epsilon+(M\Gamma/s)\exp(-\mathrm{i}\delta_R)/\sin\delta_R}\right|^{2\beta_{\mathrm{int}}}\left(1+\beta_e\frac{s-M^2}{M\Gamma}\phi\right),\qquad(17)$$

and $\beta_{\mathrm{int}}=(4\alpha/\pi)\ln\tan\frac{1}{2}\theta$.

Furthermore the finite terms $C_F^{(i)}$ can be written as

$$C_F^{(i)}=\frac{\alpha}{\pi}\{\tfrac{3}{2}(L_e+L_\mu)+4[\zeta(2)-1]\}+C_F'^{(i)},\qquad(18)$$

with $C_F'^{(i)}$ containing other $o(\alpha)$ finite terms, coming from bremsstrahlung and box diagrams, odd in the exchange $\theta\leftrightarrow\pi-\theta$, etc.

Then, after putting $\beta_{\mathrm{int}}=0$ in eqs. (15)–(17), the exponentiated factors coincide with those in eqs. (11)–(13). Furthermore the finite terms $C_F^{(i)}$ contain the $o(\beta_e,\beta_\mu)$ expansion in $\Delta_e(s)\Delta_\mu(s)$ in eqs. (11)–(13). The

[#4] This corresponds to keep $C_F^{(i)}$ to $o(\alpha)$ only.

remaining non factorizable real and virtual terms can also be included in the approach of structure functions [6]. On the other hand, the extra terms contained in eqs. (11)–(13) are of $o(\beta^2)$ – in particular the factor $-\beta_e^2\phi^2/2$ in eq. (13) arising from the expansion of $\cos\phi\beta_e$ – and $o(\epsilon\beta)$, and are not written explicitly.

From the above discussion it follows that the two formalisms give identical answers for both the exponentiated and $o(\beta)$ finite terms relative to initial and final state radiation. On the other hand they add complementary informations for the β_{int} dependence and $o(\beta^2, \epsilon\beta)$ terms, respectively.

Then, from the combined informations of eqs. (11)–(13)and (15)–(17) one can transform eq. (14) into the following form:

$$\frac{d\sigma}{d\Omega}=\frac{d\sigma_0^{(QED)}}{d\Omega}\,[C_{infra}^{(QED)}(1+\bar{C}_F^{(QED)})+C_F'^{(QED)}]+\frac{d\sigma_0^{(INT)}}{d\Omega}\,[C_{infra}^{(INT)}(1+\bar{C}_F^{(INT)})+C_F'^{(INT)}]$$

$$+\frac{d\sigma_0^{(RES)}}{d\Omega}\left[C_{infra}^{(RES)}\frac{\cos(\beta_e\phi)-\cot\delta_R\sin(\beta_e\phi)}{1+\beta_e[(s-M^2)/M\Gamma]\phi}\,(1+\bar{C}_F^{(RES)})+C_F'^{(RES)}\right], \tag{19}$$

with the definitions (15)–(17) of $C_{infra}^{(i)}$ and

$$\bar{C}_F^{(QED)}=\frac{\alpha}{\pi}\,(\Delta_e^{(1)}+\Delta_\mu^{(1)})-\beta_\mu\epsilon+\left(\frac{\alpha}{\pi}\right)^2(\Delta_e^{(2)}+\Delta_\mu^{(2)}+\Delta_e^{(1)}\Delta_\mu^{(1)})-\tfrac{1}{6}\pi^2\beta_e\beta_\mu-\tfrac{1}{4}\beta_e^2\epsilon^{1-\beta_e}, \tag{20}$$

$$\bar{C}_F^{(INT)}=\frac{\alpha}{\pi}\,(\Delta_e^{(1)}+\Delta_\mu^{(1)})-\beta_\mu\epsilon+\left(\frac{\alpha}{\pi}\right)^2(\Delta_e^{(2)}+\Delta_\mu^{(2)}+\Delta_e^{(1)}\Delta_\mu^{(1)})-\tfrac{1}{6}\pi^2\beta_e\beta_\mu$$

$$-\beta_e\frac{\cos\phi(\beta_e+1)+\tan\delta_R\sin\phi(\beta_e+1)}{\cos\phi\beta_e+\tan\delta_R\sin\phi\beta_e}\left|\frac{\epsilon}{1+\epsilon s/(M_R^2-s)}\right|$$

$$-\tfrac{1}{4}\beta_e^2\frac{\cos\phi+\tan\delta_R\sin\phi}{\cos\phi\beta_e+\tan\delta_R\sin\phi\beta_e}\left|\frac{\epsilon}{1+\epsilon s/(M_R^2-s)}\right|^{1-\beta_e}, \tag{21}$$

$$\bar{C}_F^{(RES)}=\frac{\alpha}{\pi}\,(\Delta_e^{(1)}+\Delta_\mu^{(1)})-\beta_\mu\epsilon+\left(\frac{\alpha}{\pi}\right)^2(\Delta_e^{(2)}+\Delta_\mu^{(2)}+\Delta_e^{(1)}\Delta_\mu^{(1)})-\tfrac{1}{6}\pi^2\beta_e\beta_\mu$$

$$-2\beta_e\frac{\cos\phi(\beta_e+1)-\cot\delta_R\sin\phi(\beta_e+1)}{\cos\phi\beta_e-\cot\delta_R\sin\phi\beta_e}\left|\frac{\epsilon}{1+\epsilon s/(M_R^2-s)}\right|$$

$$-\tfrac{1}{4}\beta_e^2\frac{\cos\phi-\cot\delta_R\sin\phi}{\cos\phi\beta_e-\cot\delta_R\sin\phi\beta_e}\left|\frac{\epsilon}{1+\epsilon s/(M_R^2-s)}\right|^{1-\beta_e}. \tag{22}$$

The above equations represent our final result, which describes the radiative correction factors to an accuracy better than (1%). The effect of the new terms of $o(\beta^2, \epsilon\beta)$ in eqs. (18) is shown in figs. 1, 2, where we plot the ratios $R^{(i)}=d\sigma^{(i)}[1+o(\alpha)+o(\beta\epsilon)+o(\alpha^2)]/d\sigma^{(i)}[1+o(\alpha)]$, for i=QED, INT, RES, for $\epsilon=0.01$ and $\epsilon=0.05$. Notice that the factors $C_{infra}^{(i)}$ of eqs. (15), (17) do not appear in the ratios $R^{(i)}$. We have taken the scattering angle $\theta=\frac{1}{2}\pi$ in the factors $C_F'^{(i)}$ to automatically cancel the box diagram and other non factorizable contributions.

As is clear from figs. 1, 2 the QED and INT corrections are practically constant in the resonant region to a value of about -0.01, while the RES correction is modulated essentially by the term $1-\beta_e^2\phi^2/2$, with an extra factor of o(0.01–0.02).

The extension of our results to the case of the Z line shape is straightforward. It simply corresponds to take the limit $\beta_\mu=0$, $\Delta_\mu=1$ and $\epsilon=1-4\mu^2/s$ in eq. (10). Then one simply obtains for $s\simeq M_R^2$

$$\sigma(s)\simeq\sigma_0(s)\left|\frac{M_R^2-s}{M_R^2}\right|^{\beta_e}\frac{\sin(1-\beta_e)\delta_R}{\sin\delta_R}\,\frac{\pi\beta_e}{\sin\pi\beta_e}\,\Delta_e(s), \tag{23}$$

which agrees with refs. [3,4,11] up to constant factors of $o(\beta^2)$.

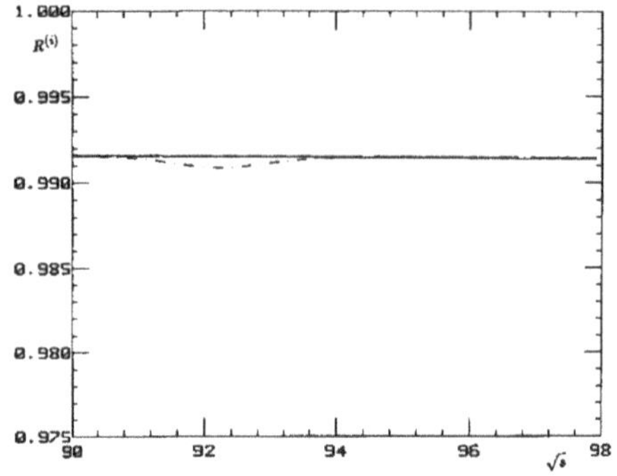

Fig. 1. Ratio $R^{(i)}=d\sigma^{(i)}[1+o(\alpha)+o(\beta\epsilon)+o(\alpha^2)]/d\sigma^{(i)}[1+o(\alpha)]$ for $e^+e^-\to\mu^+\mu^-$, with i=QED (solid line), INT (dashed line), RES (dot-dashed line) for $\epsilon=0.01$. The values of M and Γ are taken to be 92 GeV and 2.6 GeV, respectively.

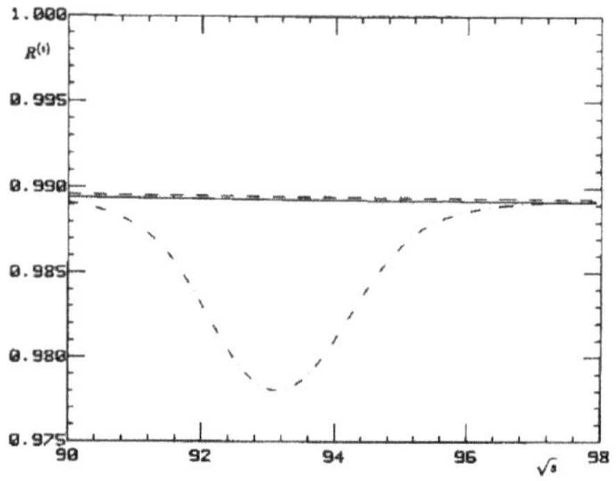

Fig. 2. Same as in fig. 1 for $\epsilon=0.05$.

To conclude, we have explicitly shown that the approach of the coherent states and that of the structure functions offer two complementary methods in QED to improve the theoretical accuracy required for precision measurements at LEP/SLC energies. They give identical results for the exponentiated and finite $o(\alpha)$ factors relative to the initial and final state radiation. Moreover, the remaining logarithmic $o(\alpha^2)$ corrections, obtained with the method of structure functions complete the information on interference effects provided by the approach of coherent states. The overall picture provides simple analytical formulae which can be easily extended to e^+e^- reactions other than $e^+e^-\to\mu^+\mu^-$.

We are grateful to G. Altarelli, O. Nicrosini and L. Trentadue for discussions.

References

[1] E.g. J. Ellis and R. Peccei, eds., Physics at LEP, CERN report CERN 86-02 (1986);
G. Alexander et al., eds., Polarization at LEP, CERN report CERN 88-06 (1988).
[2] E.g., M. Greco, Riv. Nuovo Cimento 11, no. 5 (1988).
[3] M. Greco, G. Pancheri and Y. Srivastava, Nucl. Phys. B 101 (1975) 234.
[4] M. Greco, G. Pancheri and Y. Srivastava, Nucl. Phys. B 171 (1980) 118.
[5] F.A. Berends, R. Kleiss and S. Jadach, Nucl. Phys. B 202 (1982) 63;
M. Bohm and W. Hollik, Nucl. Phys. B 204 (1982) 45.
[6] O. Nicrosini and L. Trentadue, Z. Phys. C 39 (1988) 479.
[7] F.A. Berends, W.L. van Neerven and G.J.H. Burgers, Nucl. Phys. B 297 (1988) 429.
[8] E.A. Kuraev and V.S. Fadin, Sov. J. Nucl. Phys. 41 (1985) 466;
G. Altarelli and G. Martinelli, in: Physics at LEP, eds. J. Ellis and R. Peccei, CERN report CERN 86-02 (1986);
O. Nicrosini and L. Trantadue, Phys. Lett. B 196 (1987) 551.
[9] V. Chung, Phys. Rev. B 140 (1965) 1110;
M. Greco and G. Rossi, Nuovo Cimento 50 (1967) 168.
[10] G. Curci and M. Greco, Phys. Lett. B 79 (1978) 406.
[11] R.N. Cahn, Phys. Rev. D 36 (1987) 2666.

Modern Physics Letters A
Vol. 30, No. 39 (2015) 1530031 (4 pages)

DOI: 10.1142/S0217732315300311

On the study of the Higgs properties at a muon collider

Mario Greco
Dipartimento di Matematica e Fisica, Università di Roma Tre
and
INFN, Sezione di Roma Tre, Via della Vasca Navale 84, I-00146 Rome, Italy
mario.greco@roma3.infn.it

The discovery of the Higgs particle at 125 GeV is demanding a detailed knowledge of the properties of this fundamental component of the Standard Model. To that aim various proposals of electron and muon colliders have been put forward for precision studies of the partial widths of the various decay channels. It is shown that in the case of a Higgs factory through a muon collider, sizeable radiative effects — of order of 50% — must be carefully taken into account for a precise measurement of the leptonic and total widths of the Higgs particle. Similar effects do not apply in the case of Higgs production in electron–positron colliders.

1. Introduction

The announcement of the discovery of a new scalar particle by the LHC experiments ATLAS[1] and CMS[2] has immediately triggered the question of the real nature of this particle. The determination of the spin–parity quantum numbers and the couplings to other Standard Model (SM) particles strongly suggest it to be the Higgs boson, i.e. the particle responsible for the electroweak symmetry breaking. From the available data however, it cannot be concluded yet that we have found the SM Higgs boson and not one of the scalars postulated within the possible extensions of the SM.

Therefore, a detailed study of this particle is required and various types of Higgs factories have been proposed, which can precisely determine the properties of the Higgs boson, as an important step in the future of high energy physics. In particular, the Higgs width's measurement is an essential ingredient to determine the partial width and the coupling constants to the fermions and bosons, and to this aim it is well known that lepton colliders are most appropriate for precision measurements, as happened in the past for LEP and SLC. Indeed the natural width of a 125 GeV SM Higgs boson is about 4 MeV, which is far from the precision which can be achieved at LHC.

Then circular electron–positron colliders are receiving again considerable attention. Design studies have been launched at CERN with the Future Circular

Colliders (FCC), formerly called TLEP,[3] of which an e^+-e^- collider hosted in a tunnel of about 100 km is a potential first step (FCC-ee), and in China with the Circular Electron Positron Collider (CEPC).[4] Both projects can deliver very high luminosity ($L > 10^{34}$) from the Z peak to HZ threshold. The Higgs bosons are produced mainly through Higgs-strahlung, at center-of-mass energy of about 250 GeV, and therefore the Higgs boson is produced in association with a Z boson. Then the Higgs signal can be tagged with the Z boson decay, especially if the Z boson decays into a pair of leptons. The HZZ coupling can be inferred in this way, but it seems to us that the Higgs width is not achieved so easily.

On the other hand the idea of a muon collider with $L > 10^{32}$ seems much more appealing.[5-7] The collider ring is much smaller, with $R \sim 50$ m, and here the direct H^0 cross-section is greatly enhanced with respect to e^+-e^- since the s-channel coupling is proportional to the square of the initial lepton mass. In analogy to the case of the Z^0 at LEP/SLC, the production of a single H^0 scalar in the s-state offers unique conditions of experimentation because the H^0 has a very narrow width and most of its decay channels can be directly compared to the SM predictions with a very high accuracy. The drawback of this project however is that a very powerful "muon cooling" is needed.[a] A practical realization of a demonstrator of a muon cooled Higgs factory has been recently suggested by Rubbia.[7]

In this note we show that in the case of a muon collider very important QED radiative corrections change drastically the lowest order Higgs production results, and must be taken in full account for an accurate description of the Higgs total and partial widths, as well as the various H^0 couplings. Such type of effects do not apply in the case of electron positron colliders, when the Higgs bosons are produced mainly through Higgs-strahlung in association with a Z boson. The origin of these radiative effects is well known and has been discussed in very great detail in the case of collisions of electrons and positrons with production of narrow resonances in the s-channel like the J/Psi[8] and the Z boson.[9] Namely a correction factor $\propto (\Gamma/M)^{(4\alpha/\pi)\log(2E/m)}$ modifies the lowest order cross-section, where M and Γ are the mass and width of the s-channel resonance, $W = 2E$ is the total initial energy and m is the initial lepton mass. Physically this is understood by saying that the width provides a natural cutoff in damping the energy loss for radiation in the initial state. To be more specific, defining

$$\beta_i = \frac{4\alpha}{\pi}\left[\log\frac{W}{m_i} - \frac{1}{2}\right], \tag{1}$$

where m_i is the initial lepton mass,

$$y = W - M\,,$$

$$\tan\delta_R(W) = \frac{1}{2}\Gamma/(-y)$$

[a]For a detailed discussion and a complete set of references, see for example, Ref. 7.

then the infrared factor $C^{\rm res}_{\rm infra}$ due to the soft radiation emitted from the initial charged leptons is given by[8]

$$C^{\rm res}_{\rm infra} = \left(\frac{y^2 + (\Gamma/2)^2}{(M/2)^2}\right)^{\beta_i/2}\left[1 + \beta_i \frac{y}{\Gamma/2}\delta_R\right], \tag{2}$$

so that the observed resonant cross-section can be written as

$$\sigma^c = C^{\rm res}_{\rm infra}\sigma_{\rm res}(1 + C^{\rm res}_F)\,. \tag{3}$$

In the above equation $\sigma_{\rm res}$ is the Born resonant cross-section (of Breit–Wigner form) and $C^{\rm res}_F$ is a finite standard correction of order α which we will neglect in the following. In the case of Higgs production at a muon collider with $W = 2E = 125$ GeV the factor $\beta_i = 0.061$ and at the resonance ($y = 0$) the factor $C^{\rm res}_{\rm infra} = (\Gamma/M)^{\beta_i} = 0.53$, assuming the Higgs width $\Gamma = 4$ MeV, which gives a substantial reduction of the Born cross-section and therefore can mimic a smaller initial (and/or final) partial decay width of the Higgs.

As it is well known, since the produced resonance is quite narrow, one has to integrate over the machine resolution, which is assumed to be

$$G(W' - W) = \frac{1}{\sqrt{2\pi}\sigma}e^{-(W'-W)^2/(2\sigma^2)}\,, \tag{4}$$

where σ is the machine dispersion, such that $(\Delta W)_{\rm FWHM} = 2.3548\sigma$. Then the experimentally observed cross-section into a final state $|f\rangle$ is given by

$$\tilde{\sigma}(W) = \int G(W' - W)dW'\sigma(W')\,, \tag{5}$$

where

$$\sigma(W') = \frac{4\pi}{W'^2}\frac{\Gamma_i\Gamma_f}{\Gamma^2}\sin^2\delta_R(W')\left\{\frac{\Gamma}{W'\sin\delta_R(W')}\right\}^{\beta_i}$$
$$\times(1 - \beta_i\delta_R\cot\delta_R)(1 + C^{\rm res}_F)\,. \tag{6}$$

By integrating the last equation, a useful formula for the observed cross-section at the peak is[8]

$$\tilde{\sigma}(M) = \frac{2\pi^2\Gamma_i\Gamma_f}{\sqrt{2\pi}\sigma M^2\Gamma}\left(\frac{\Gamma}{M}\right)^{\beta_i}e^{\left(\frac{\Gamma}{2\sqrt{2}\sigma}\right)^2}$$
$$\times\left\{\mathrm{ercf}\left(\frac{\Gamma}{2\sqrt{2}\sigma}\right) + \frac{1}{2}\beta_i E_1\left(\frac{\Gamma^2}{8\sigma^2}\right)\right\}(1 + C^{\rm res}_F)\,, \tag{7}$$

where the second term in the square bracket represents the contribution from the radiative tail.

Numerically, and neglecting the contribution from $C^{\rm res}_F$, the overall radiative correction factor C to the Born cross-section at the peak is $C = 0.47$, 0.37, 0.30, 0.25 for $\sigma = 1$ MeV, 2 MeV, 3 MeV, 4 MeV, respectively. This result shows once again the importance of the radiative effects for a precision measurement of the Higgs couplings.

Our analytical approach, with an accuracy of order of a few %, gives a good estimate of the radiation effects expected for the Higgs signal around the resonance. On the other hand, the background final states, as b quark–antiquark pairs and W pairs, coming from the direct Z production and decay, estimated for example in Ref. 10, are not substantially affected from the initial state radiation. Of course a more precise numerical evaluation of the ratio Signal/Background is in order, when the muon collider will be finalized.

To conclude, we have shown that in the case of a Higgs factory through a muon collider, sizeable radiative effects — of order of 50% — must be carefully taken into account for a precise measurement of the leptonic and total widths of the Higgs particle. Similar effects do not apply in the case of Higgs production in electron–positron colliders.

References

1. ATLAS Collab. (G. Aad *et al.*), *Phys. Lett. B* **716**, 1 (2012).
2. CMS Collab. (S. Chatrchyan *et al.*), *Phys. Lett. B* **716**, 30 (2012).
3. M. Koratzinos *et al.*, arXiv:1305.6498.
4. D. Wang and J. Gao, Study on Beijing Higgs Factory (BHF) and Beijing Hadron Collider (BHC), IHEP-AC-LC-Note 2012-012.
5. D. B. Cline, *Nucl. Instrum. Meth. A* **350**, 24 (1994).
6. V. D. Barger, M. S. Berger, J. F. Gunion and T. Han, *Phys. Rep.* **286**, 1 (1997).
7. C. Rubbia, arXiv:1308.6612.
8. M. Greco, G. Pancheri-Srivastava and Y. Srivastava, *Nucl. Phys. B* **101**, 234 (1975).
9. M. Greco, G. Pancheri-Srivastava and Y. Srivastava, *Nucl. Phys. B* **171**, 118 (1980) [Erratum-*ibid.* **197**, 543 (1982)].
10. C. Han and Z. Liu, arXiv:1210.7803v2.

Physics Letters B 763 (2016) 409–415

Contents lists available at ScienceDirect

Physics Letters B

www.elsevier.com/locate/physletb

ISR effects for resonant Higgs production at future lepton colliders

Mario Greco [a], Tao Han [b,c,d], Zhen Liu [e,*]

[a] Dipartimento di Matematica e Fisica, Universita' di Roma Tre, INFN, Sezione di Roma Tre, Via della Vasca Navale 84, I-00146 Rome, Italy
[b] Pittsburgh Particle Physics, Astrophysics, and Cosmology Center, Department of Physics and Astronomy, University of Pittsburgh, 3941 O'Hara St., Pittsburgh, PA 15260, USA
[c] Department of Physics, Tsinghua University, Beijing, 100086, China
[d] Collaborative Innovation Center of Quantum Matter, Beijing, China
[e] Theoretical Physics Department, Fermi National Accelerator Laboratory, Batavia, IL, 60510, USA

ARTICLE INFO

Article history:
Received 14 July 2016
Received in revised form 28 October 2016
Accepted 29 October 2016
Available online 4 November 2016
Editor: L. Rolandi

Keywords:
Higgs boson width
Lepton collider
Initial state radiation

ABSTRACT

We study the effects of the initial state radiation on the s-channel Higgs boson resonant production at $\mu^+\mu^-$ and e^+e^- colliders by convoluting with the beam energy spread profile of the collider and the Breit–Wigner resonance profile of the signal. We assess their impact on both the Higgs signal and SM backgrounds for the leading decay channels $h \to b\bar{b}$, WW^*. Our study improves the existing analyses of the proposed future resonant Higgs factories and provides further guidance for the accelerator designs with respect to the physical goals.

1. Introduction

The Higgs boson discovery at the LHC in 2012 [1,2] has opened a new era of particle physics: It is the first elementary scalar particle ever observed in Nature and its properties thus need to be thoroughly scrutinized. Future lepton collider Higgs factories [3–12] are proposed to study the Higgs boson properties to great accuracies because of the much more favorable experimental environment than that at hadron colliders [13]. Amongst many candidates of Higgs factories, the possibility of s-channel resonant production is especially important. The muon collider Higgs factory could produce the Higgs boson in the s-channel and perform an energy scan to map out the Higgs resonance line shape at tens of MeV level [3–5]. This approach would provide the most direct measurement of the Higgs boson total width and the Yukawa coupling to muons. The clean environment of the lepton colliders with a large number of Higgs bosons produced also enables precision measurements for many exclusive decays of the Higgs boson. More recently the possibility of an ultra high luminosity electron–positron collider for the Higgs resonant production has been proposed [14], providing a possible opportunity to observe the Higgs signal and thus the determination of the Yukawa coupling to electrons — so far the only conceivable measurement of the Higgs coupling to the first generation of fermions [15].

Due to the narrow width of the Higgs boson, about 4.07 MeV [16] as predicted by the Standard Model (SM), it would be extremely demanding for the collider energy resolution to reach a similar value in order to adequately study the physical width. This has been quantified in the literature by convoluting the Breit–Wigner resonance for the Higgs signal and the Gaussian distribution for the profile of beam energy spread (BES) [3,5]. It is also known that, the Initial State Radiation (ISR) of the QED effect would degrade the peak luminosity of a lepton collider [4]. The impact on a muon collider has been recently emphasized [17,18], and the effects would be notably stronger for an e^+e^- collider because of a lighter electron mass. In this work, we study all the effects coherently for a few representative choices of the BES and different approximations for the ISR. We assess their impact in different scenarios on both the Higgs signal and SM background. Our study improves the existing analyses of the proposed future resonant Higgs factories and provides further guidance for the target accelerator designs with respect to the physical goals.

The work is organized as follows. In Section 2 we present the formulation and parameterization of the BES and ISR effects. In Section 3 we quantify their effects on the Higgs boson signal and the SM backgrounds on a muon collider, and study the observability for the Higgs signal at an e^+e^- collider. We conclude in

* Corresponding author.
E-mail addresses: mario.greco@roma3.infn.it (M. Greco), than@pitt.edu (T. Han), zliu2@fnal.gov (Z. Liu).

http://dx.doi.org/10.1016/j.physletb.2016.10.078
0370-2693/

Section 4. Some analytical formulas adopted in our calculations for the ISR are listed in Appendix A.

2. BES and ISR parameterization in resonant Higgs boson production

2.1. BES parameterization

As studied to a great detail in the literature [3–5], the muon collider energy resolution is critically important to study the Higgs width and interactions due to the very narrow width of the Higgs boson. The observable cross section is given by the convolution of the energy distribution delivered by the collider. We assume that the lepton collider c.m. energy ($\sqrt{s}$) has a flux L distribution

$$\frac{dL(\sqrt{s})}{d\sqrt{\hat{s}}} = \frac{1}{\sqrt{2\pi}\,\Delta}\exp[\frac{-(\sqrt{\hat{s}}-\sqrt{s})^2}{2\Delta^2}],$$

with a Gaussian energy spread $\Delta = R\sqrt{s}/\sqrt{2}$, where R is the percentage beam energy resolution, then the effective cross section is

$$\sigma_{\text{eff}}(s) = \int d\sqrt{\hat{s}}\,\frac{dL(\sqrt{s})}{d\sqrt{\hat{s}}}\,\sigma(\ell^+\ell^- \to h \to X)(\hat{s}) \qquad (2.1)$$
$$\propto \begin{cases} \Gamma_h^2 B/[(s-m_h^2)^2+\Gamma_h^2 m_h^2] & (\Delta \ll \Gamma_h), \\ B\exp[\frac{-(m_h-\sqrt{s})^2}{2\Delta^2}](\frac{\Gamma_h}{\Delta})/m_h^2 & (\Delta \gg \Gamma_h). \end{cases}$$

The interaction strength B is proportional to the Higgs coupling squared and governs the overall normalization for the Higgs production rate. For $\Delta \ll \Gamma_h$, the line shape of a Breit–Wigner resonance can be mapped out by scanning over the energy $\sqrt{s}$ as given in the first equation. For $\Delta \gg \Gamma_h$ on the other hand, the physical line shape is smeared out by the Gaussian distribution of the beam energy spread and the signal rate will be determined by the overlap of the Breit–Wigner and the luminosity distributions, as seen in the second equation above. Our results for the Higgs line-shape will be discussed in detail in the next section. In addition to the Higgs signal, an important issue of phenomenological interest is the question of the expected background in the various Higgs decay channels. That is mainly related to the tail of the Z-boson produced in the lepton annihilation. This issue has already been a subject of study in Ref. [5], but has to be reviewed in the light of the corrections introduced by the radiative effects. Again those corrections will be calculated using the theoretical approach discussed above. Then the results for the Signal/Background ratio for various final state configurations will be also discussed in the next section.

2.2. ISR parameterization

Multiple soft photon radiation in the initial state (ISR) is an important effect in high energy lepton collisions [19]. In particular, when a narrow resonance is produced in an s-channel annihilation, the ISR effect becomes more significant. The first prominent example of such effects was the historical observation of J/ψ production in e^+e^- annihilation [20]. The origin of the ISR effect is well-known and was earlier discussed in great detail near the J/ψ peak [21], and later for the case of the Z-boson production [22]. Qualitatively, a modification factor to the lowest order cross section can be expressed by

$$\kappa \propto (\frac{\Gamma}{M})^{\frac{4\alpha}{\pi}\log(\frac{\sqrt{\hat{s}}}{m})}$$

where M and Γ are the mass and width of the s-channel resonance, $\sqrt{\hat{s}}$ is the c.m. energy in the partonic collision, and m is the beam lepton mass. Physically this implies that the width provides a natural cut-off in damping the energy loss for radiation in the initial state. Very precise calculation techniques for these QED effects have been developed for LEP experiments, where in addition to multi-photon radiation finite corrections have been added, by including, at the least, up to two-loop effects [23,24]. In the case of muon colliders, in particular for Higgs boson production studies, those effects were not emphasized sufficiently in the past, and only recently their importance has been pointed out [17,18] for the experimental study of the Higgs line-shape as well as for the machine design of the initial BES. In particular the estimates of the reduction factors of the Higgs production cross sections, of order of 50% or more, depending upon the machine energy spread, given in Ref. [17], have been confirmed in Ref. [18], with the Higgs line-shape explicitly shown.

We will make use of the general formalism of the electron (muon) structure functions, first introduced in Ref. [25], and later improved for LEP experiments, which is well suited for the numerical calculations of the various distributions of phenomenological interest. Our goal is to produce integrated cross sections to an accuracy of $O(1\%)$ which could be used as a reference in current studies of lepton Higgs factories. For the sake of completeness, we will compare various levels of the approximation which can be found in the literature for the lepton structure functions. As a first calculation technique we will use the approach of Ref. [26], where in addition to the exponentiated effect from multi-photon radiation, finite terms have been included up to the second order. The results will be compared with the approach discussed in Ref. [27] – to various levels of approximation – and explicitly adopted in Ref. [18] for the Higgs line-shape.

The initial state radiation (ISR) effect collectively can be expressed with different levels of sophistication. We define the probability distribution function $f_{\ell\ell}^{\text{ISR}}(x)$ for the hard collision energy $x\sqrt{\hat{s}}$, and hard collision cross section

$$\sigma(\ell^+\ell^- \to h \to X)(\hat{s}) = \int dx\, f_{\ell\ell}^{\text{ISR}}(x;\hat{s})\hat{\sigma}(\ell^+\ell^- \to h \to X)(x^2\hat{s}), \qquad (2.2)$$

where x is the fraction of the c.m. energy at the hard collision with respect to the beam energy before the collision. We list several commonly used analytical formulas for the ISR parameterization under different approximations in Appendix A.

In particular, for completeness and the convenience of the readers, we summarize here the various approximations in the literature. In Ref. [25] only some $O(\alpha)$ terms are included in addition to the exponentiated soft radiation term. In Ref. [27] the various approximations contain: (a) the soft exponentiated term only; (b) adds to (a) the full $O(\alpha)$ terms; and (c) adds to (b) the relevant $O(\alpha^2)$ terms. Finally Ref. [26] contains the full exponentiated term and the complete $O(\alpha)$ and $O(\alpha^2)$ terms. The explicit expressions of various approximations are provided in Appendix A.

In Fig. 1, we show those energy distributions versus the energy fraction x with the ISR effects in various approximations, for $\mu^+\mu^-$ (left panel) and e^+e^- (right panel) initial beams for a c.m. energy $\sqrt{s} = 125$ GeV. We can see that the widely used Kuraev–Fadin [25] approximation (lower red curves) is more steep in falling comparing to other improved calculations. The approach of Jadach–Ward–Was [27] improves the approximations at different orders and complexities as listed in Appendix A. We see that their benchmark choice (a) leads to a much larger radiation tail (upper blue curves). On the other hand, the more sophisticated approximations of Ref. [27] (b) and (c) (middle orange and green curves) agree much better with earlier works by Nicrosini and Trentadue [26] (shown as the black curves). This comparison signifies the importance of the proper treatment in evaluating the ISR

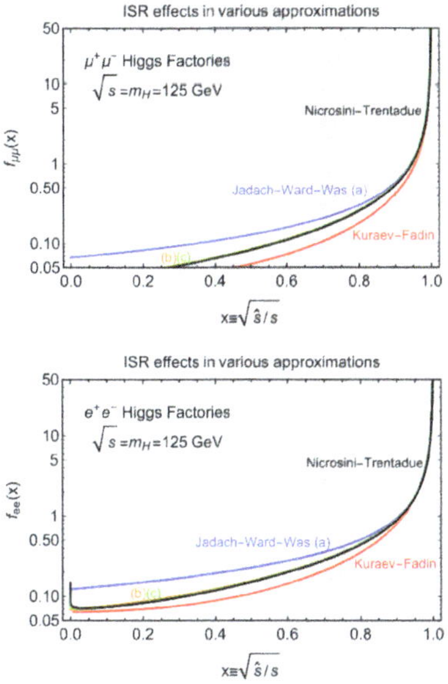

Fig. 1. The beam energy distribution as a function of the energy fraction x from ISR effects in various approximations, for $\mu^+\mu^-$ (upper panel) and e^+e^- (lower panel) initial states for 125 GeV center of mass energy. The shaded brown bands correspond to the collision energy near the Z-boson mass in the $\pm 2\Gamma_Z$ window. (For interpretation of the references to color in this figure, the reader is referred to the web version of this article.)

effects. Henceforth, we will restrict to the approaches of Ref. [26] and the choices (b) and (c) of Ref. [27], which will lead us to our final calculated cross sections with the estimate of the theoretical accuracy. In particular we have found that at the level of the structure functions $f_{\ell\ell}^{\rm ISR}(x;\hat{s})$ the relative difference between the formalism (b) by Jadach–Ward–Was [27] and that of Nicrosini–Trentadue [26] is at 1∼2% level and 4∼5% level for large and small values of x, respectively. However the relative difference between the calculated Higgs cross sections is smaller because in the convolution only high-x matter. The $\mu^+\mu^-$ case has of course a better agreement comparing to the e^+e^- case because of the smaller radiative effects of muons. In particular, in the muon case, we have checked that the relative difference between the formalism (b) by Jadach–Ward–Was [27] and that of Nicrosini–Trentadue [26] for the final cross sections is within 1%, which is the estimate of the theoretical accuracy of our results. On the other hand the cross sections difference between the formalisms (b) and (c) by Jadach–Ward–Was [27] is O(0.1%). As we are performing a convolution of the ISR effect over BES effect and then over the Breit–Wigner profile for a scan, the computational accessibility is important here. We have found that the evaluation time for the structure functions alone in the formalism of Nicrosini–Trentadue [26] is about 5 times larger than the formalism (b) by Jadach–Ward–Was [27], at any value of x. Consequently, we choose formalism (b) as a balanced formalism between speed and accuracy. Our results, as stated above, will have a theoretical accuracy of O(1%).

As a consequence of the ISR, a very significant phenomenon is the "radiative return" to a lower mass resonance. Despite the beam collision energy is above a resonance mass, after ISR radiation, the hard collision center of mass energy "returns" to the resonance mass and hit the Breit–Wigner enhancement again. This mechanism can be utilized to effectively producing lighter resonances without scanning the beam energy [28,29]. In Fig. 1, we shade the region in brown color for the x values corresponding to the $\pm 2\Gamma_Z$ window near Z-boson mass for a 125 GeV lepton collider. The rate in this window predicts the amount of "radiative return" Z bosons produced, which constitutes a large background for Higgs studies. Once again, we can see that different parameterizations of the ISR effects yield significantly different amount of "radiative return" Z production rate.

Table 1
Effective cross sections in $\mu^+\mu^-$ (upper panel) collision in units of pb and e^+e^- (lower panel) collision in units of fb at the resonance $\sqrt{s} = m_h = 125$ GeV, with Breit–Wigner resonance profile alone, with ISR alone (Jadach–Ward–Was (b)), with BES alone for two choices of beam energy resolutions, and both the BES and ISR effects included.

σ(BW)	ISR alone	R (%)	BES alone	BES+ISR
$\mu^+\mu^-$: 71 **pb**	37	0.01	17	10
		0.003	41	22
e^+e^-: 1.7 **fb**	0.50	0.04	0.12	0.048
		0.01	0.41	0.15

3. Numerical studies on the ISR and beam effects

The ISR effects, as discussed in details in previous sections, are very important and inevitable at future lepton collider resonant Higgs factories. The ISR effects need to be convoluted with the finite BES as expressed in Eq. (2.1). We evaluate numerically their importance in the Higgs boson property measurements in this section.

3.1. The case for the muon collider

The muon collider Higgs factory features a line-shape scan of the Higgs boson, enables a simultaneous measurement of the Higgs boson mass, width and muon Yukawa at unprecedented precision [3–5]. The inclusion of the ISR effects make the prediction more robust.

In Table 1 we show the reduction effects for the resonance production of the SM Higgs boson at 125 GeV for a muon collider (upper panel) including BES and ISR. The resonance production rate is reduced by a factor of 1.9 with the inclusion of ISR effect with the parameterization of Jadach–Ward–Was (b). Independently, the production rate would be reduced by factors of 4.2 and 1.7 for beam spread of 0.01% and 0.003% respectively.[1] The total reduction after the convolution of the beam spread and the ISR effect is 7.1 and 3.2 for the two beam spread scenarios, respectively.

To illustrate the resulting line-shape we show in Fig. 2 (left panel for a $\mu^+\mu^-$ collider) for various setups of our evaluation. We show the sharp Breit–Wigner resonance in solid blue lines. The BES will broaden the resonance line-shape with a lower peak value and higher off-resonance cross sections, as illustrated by the green curves. The solid lines and dashed lines represent the narrow and wide BES of 0.01% and 0.003%, respectively. The ISR effect is asymmetric below and above the resonant mass, because it only reduces the collision energy by emitting photons, shown in the orange curve. In regions 10 MeV above the Higgs mass, the ISR

[1] In comparison with the cross sections considering beam energy spread in our initial study [5], some small numerical differences are generated due to a different choice of the Higgs boson mass of 125 GeV instead of 126 GeV and correspondingly the different branching fractions and total widths.

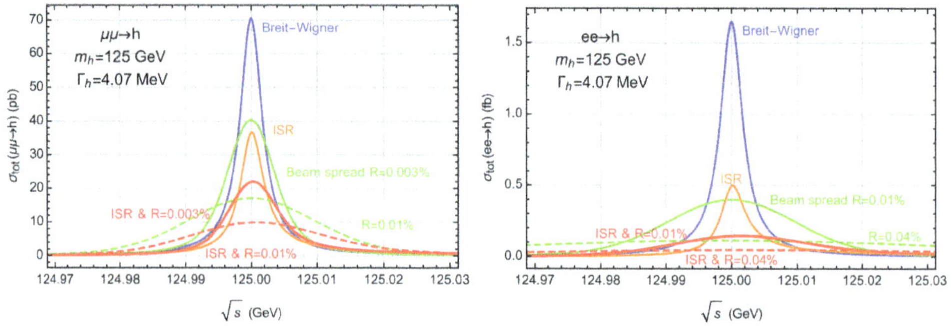

Fig. 2. The line shapes of the resonances production of the SM Higgs boson as a function of the beam energy $\sqrt{s}$ at a $\mu^+\mu^-$ collider (left panel) and an e^+e^- collider (right panel). The blue curve is the Breit-Wigner resonance line shape. The orange line shape includes the ISR effect alone for Jadach-Ward-Was (b). The green curves include the BES only with two different energy spreads. The red line shapes take into account all the Breit-Wigner resonance, ISR effect and BES in solid and dashed lines, respectively. (For interpretation of the references to color in this figure, the reader is referred to the web version of this article.)

effect increases the production rate via "radiative return" mechanism. Still, the overall effect is the reduction of on-shell rate as clearly indicated in the plot. In red lines we show the line shapes of the Higgs boson with both the BES and the ISR effect. We can see the resulting line shape is not merely a product of two effect but rather complex convolution, justifying necessity of our numerical evaluation.

Having understood the ISR and BES effects on the signal production rates and line shapes, we now proceed to understand the effect on the background. For the muon collider study, the main search channels for the Higgs boson will be the exclusive mode of $b\bar{b}$ and WW^*. For the $b\bar{b}$ final state the main background is from the off-shell Z/γ s-channel production. The ISR and BES effects barely change the rate from such off-shell process. However, the ISR effect does increase the on-shell $Z \to b\bar{b}$ background through the "radiative return" mechanism. Our numerical study shows that the "radiative return" of the Z boson to $b\bar{b}$ increase the inclusive $b\bar{b}$ background by a factor of seven. Since we understand that the increase of the background is dominantly from the on-shell Z boson, the new background rates after imposing a $b\bar{b}$ invariant mass cut of 95, 100, 110 GeV, change to 17, 20, 25 pb, respectively. Given the finite resolution of the b-jet energy reconstruction, we propose an invariant mass cut of the $b\bar{b}$ system of 100 GeV, which leads to around 20% increase in such background comparing to the tree-level estimate. So far we have suggested the invariant mass cut for the $b\bar{b}$ pair, as an example of discrimination from the background. One could also foresee a cut on the angle between the two b-jets, which could be measured more precisely than the invariant mass.[2]

Beyond the $b\bar{b}$ final state, another major channel for muon collider Higgs physics is the WW^* channel. This channel enjoys little (irreducible) background form the SM process. The ISR effect introduces no "radiative return" for such process. Consequently, the background rate does not change from the tree-level estimate. We summarize in Table 2 the on-shell Higgs production rate and background rate in these two leading channels with the inclusion of the ISR and BES effects. We can see from the table that at the muon collider Higgs factory, the signal background ratio is pretty large and the observability is simply dominated by the statistics. The "radiative return" from the ISR effect, however, does impact several other Higgs decay channel search more. For example, searches of Higgs rare decay of $h \to Z\gamma$, Higgs decay of $h \to ZZ^*$ with

Table 2
Signal and background effective cross sections at the resonance $\sqrt{s} = m_h = 125$ GeV at a $\mu^+\mu^-$ collider (upper panel, in pb) and an e^+e^- collider (lower panel, in ab) for two choices of beam energy resolutions R and two leading decay channels with ISR effects taken into account, with the SM branching fractions $\mathrm{Br}_{b\bar{b}} = 58\%$ and $\mathrm{Br}_{WW^*} = 21\%$. For the $b\bar{b}$ background, a conservative cut on the $b\bar{b}$ invariant mass to be greater than 100 GeV is applied.

R (%)	$\mu^+\mu^- \to h$	$h \to b\bar{b}$		$h \to WW^*$	
	$\sigma_{\rm eff}$ (**pb**)	σ_{Sig}	σ_{Bkg}	σ_{Sig}	σ_{Bkg}
0.01	10	5.6	20	2.1	0.051
0.003	22	12		4.6	
R (%)	$e^+e^- \to h$	$h \to b\bar{b}$		$h \to WW^*$	
	$\sigma_{\rm eff}$ (**ab**)	σ_{Sig}	S/B	σ_{Sig}	S/B
0.04	48	27	$\mathcal{O}(10^{-6})$	10	$\mathcal{O}(10^{-3})$
0.01	150	81		31	

Table 3
Fitting accuracies for one standard deviation of Γ_h, B and m_h of the SM Higgs with the scanning scheme for two representative luminosities per step and two benchmark beam energy spread parameters.

$\Gamma_h = 4.07$ MeV	L_{step} (fb^{-1})	$\delta\Gamma_h$ (MeV)	δB	δm_h (MeV)
$R = 0.01\%$	0.05	0.79	3.0%	0.36
	0.2	0.39	1.1%	0.18
$R = 0.003\%$	0.05	0.30	2.5%	0.14
	0.2	0.14	0.8%	0.07

$Z^* \to \nu\bar{\nu}$, etc are facing more challenges and new selection cuts need to be designed and applied.

Finally, we perform a study on the potential precision on the Higgs properties at a future muon collider through a lineshape scan. We follow the benchmarks, statistical treatment and procedure defined in Ref. [5], where a 21 steps scan in the mass window of ± 30 MeV around the Higgs mass with equal integrated luminosities.[3] A fit to the result of such lineshape scan can simultaneously determine the Higgs total width Γ_h, the Higgs mass m_h and interaction strength B with great precision. The interaction strength B can be directly translated into the Higgs muon Yukawa after fixing the decay branching fractions or performing a global fit. We tabulate the projected precisions on these quantities in Table 3 for the two benchmark BES values of $R = 0.01\%$ and

[2] We thank the Editor Gigi Rolandi for suggesting this discrimination procedure.

[3] The Higgs mass may not known to the ± 30 MeV level by the time of the muon collider, and a pre-scan stage to determine the Higgs mass will be required [30].

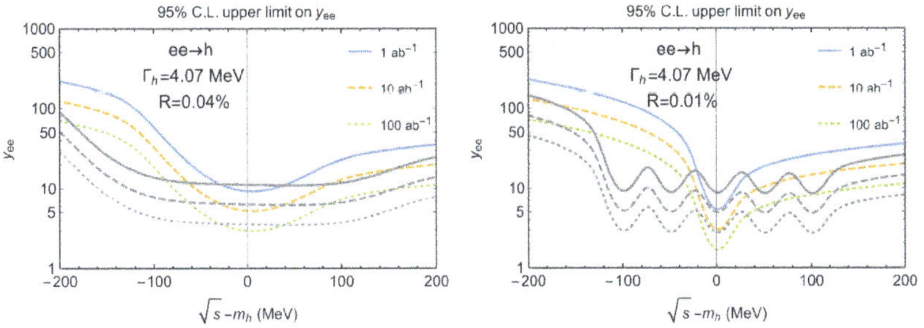

Fig. 3. The projected 95% C.L. upper limit on the Higgs electron Yukawa coupling at an electron–positron resonant Higgs factory with different integrated luminosities for two operational strategies. We consider two benchmark beam energy resolutions of 0.04% and 0.01% in the left panel and the right panel, respectively. (For interpretation of the references to color in this figure, the reader is referred to the web version of this article.)

$R = 0.003\%$ and two benchmark integrated luminosities per scan step of 0.05 fb^{-1} and 0.2 fb^{-1}. For the case of optimistic BES of $R = 0.003\%$, we find that all related Higgs properties can be determined to great precision, including the Higgs width to 0.14 MeV, the Higgs mass to 0.07 MeV and the interaction strength B to 0.8% with 0.2 fb^{-1} per scan step. For a lower statistics of 0.05 fb^{-1} per scan step, the projected precision are basically doubled, following the statistical dominance argument. For the conservative BES of $R = 0.01\%$ with different luminosities, the achievable precision is roughly the result for $R = 0.003\%$ doubled, in more detail the precision on Higgs width and interaction strength are still great at percent level while the precision on Higgs mass remains at sub MeV level.

3.2. The case for the electron–positron collider

The case for the electron–positron collider for a resonant production of the SM Higgs boson has a rather different physics purpose. Unlike the muon collider case for a precision measurement of many crucial properties of the Higgs boson, including the width, mass, muon Yukawa coupling with unprecedented precision, the most important physics goal of an electron–positron collider at 125 GeV is to constrain the electron Yukawa coupling. This would certainly be a first $O(1) \sim O(10)$ level test on this first generation Yukawa coupling. The potential to probe this first generation Yukawa will provide important implications for a broad class of flavor models.

In electron–positron collisions, ISR effect is significantly larger because the radiation is inversely proportional to lepton mass squared. The ISR effect is further amplified by the beamstrahlung due to the demand of a high instantaneous luminosity. These lead to a broadening of the beam energy distribution. The on-peak cross section is more notably reduced than that at a muon collider. In Table 1 (lower panel), we show the on-resonance production rate reduction for the SM Higgs boson at 125 GeV. The on-resonance production rate is reduced by a factor of 3.4 with the inclusion of ISR effect. The achievable beam parameters for the possible electron–positron resonant Higgs factory is not clear so far. For the sake of demonstration, we choose two benchmark cases of the BES: $R = 0.04\%$, which is running design for the FCC-ee at Z-pole; and an improved design $R = 0.01\%$ for possible future developments [31]. The production reduction factors are 14 and 4.2 for those two beam spreads, respectively. The total reduction after the convolution of the beam spread and the ISR effect is 35 and 11 for the two beam spread scenarios, respectively as shown in the last column in the table.

To illustrate the resulting line-shape we show in Fig. 2 (right panel for an e^+e^- collider) for various setups of our evaluation. The sharp Breit-Wigner resonance is shown by the solid blue line. The BES will broaden the resonance line-shape with a lower peak value and higher off-resonance cross sections, as illustrated in the green curves, with the solid line and dashed line representing the two BES parameterizations of 0.04% and 0.01%, respectively. The resulting line-shape features are very similar to the case of a muon collider as shown in the left panel, but with larger reduction factors from both the BES and ISR effects.

In Table 2 at the lower panel, similar to the muon collider case, we list signal rates for the two leading decay channels of the SM Higgs boson for the electron–positron resonant Higgs factory. The signal rates for the two leading Higgs decay channels are all at tens of attobarn level. The background rates are the same as listed in the muon collider case in Table 2 and we hence list the signal background ratio S/B instead. We observe that the S/B for the $h \to b\bar{b}$ process is quite small $\mathcal{O}(10^{-6})$ and this channel will not be contributing much to the Higgs physics. Next, the $h \to WW^*$ will be the leading channel for the consideration, if assuming that the systematics can be controlled at $\mathcal{O}(10^{-3})$ level.

In Fig. 3 we show the projected 95% C.L. upper limit on the Higgs-electron Yukawa coupling (normalized to the SM value) with various collider running scenario and search strategies as a function of the difference between the beam energy and the Higgs pole mass ($\sqrt{s} - m_h$) based on the exclusive channel of $e^+e^- \to h \to WW^*$. We demonstrate two strategies here, one fixed energy for the full integrated luminosity (shown in colored lines) and one five-step scan with 50 MeV intervals around the Higgs mass (shown in gray lines) with equal shares of the total integrated luminosities. The results for the two benchmark case of BES $R = 0.04\%$ and $R = 0.01\%$ are displayed in the left panel and right panel, respectively. In both panels, the exclusion limits are shown assuming null observation beyond the SM expectation, and from up to down, the solid, dashed and dotted lines represent the integrated luminosity of 1, 10, 100 ab^{-1}, respectively.

As for the running strategy of sitting on a single energy for the total integrated luminosity, the upper limit on the electron Yukawa can reach 3, 5 and 8 times the SM value for $R = 0.04\%$ (left panel) and 1.7, 3, 5 times the SM value for $R = 0.01\%$ (right panel) with an integrated luminosity of 1, 10, 100 ab^{-1}, respectively, if the beam energy is tuned right at the Higgs mass. If the beam energy is set to be around 60 MeV (30 MeV) above the Higgs pole mass, the upper limit on the electron Yukawa is doubled. The asymmetric behavior of Higgs line-shape generated by the ISR effect appears here as the exclusion limits degrades much faster when the beam energy is below the Higgs mass than when above. However, we may not have *a priori* precise knowledge of the Higgs mass. We thus demonstrate an alternative strategy of a 5 step scanning with an interval of 50 MeV around the Higgs mass in gray lines. We can see in the gray curves that this strategy provides a relative flat sensitivity across the ±100 MeV range for BES $R = 0.04\%$, yielding an upper bound of around 3.5 ∼ 4 times the SM electron Yukawa with 100 ab^{-1} integrated luminosity. For BES $R = 0.01\%$ The wavy structure in the exclusion limits indicates the divide of the scanning steps is not fine enough, and the exclusion at 100 ab^{-1} various between 3 to 5 times the SM electron Yukawa. We conclude that for a single energy run, the better beam energy resolution than $R = 0.04\%$ is not advantageous unless a knowledge on the Higgs mass precision of around 10 MeV is available before choosing the beam energy. If the Higgs mass is known to a level of ±50 MeV, a multistep scan can provide a rather uniform exclusion limits in the ±100 MeV window of the Higgs mass, reaching around 3 times the SM electron Yukawa, and in this case better BES simply means more scanning steps in this mass window.

4. Conclusion

We studied the effects from the initial state radiation and beam energy spread coherently for lepton colliders for the narrow Higgs boson production in the s-channel. We presented a few representative choices of the BES and different approximations for the ISR. We quantify their impact in different scenarios for both the Higgs signal and SM background. We found that

- The BES effect is potentially the leading factor for the resonant signal identification, and it alone reduces the on resonance Higgs production cross section by a factor of 1.7 (4.2) for a muon collider with $R = 0.003\%$ ($R = 0.01\%$), and by a factor of 4.2 (14) for an electron–positron collider with $R = 0.01\%$ ($R = 0.04\%$), as shown in Table 1.
- The ISR effect alone reduces the on-resonance Higgs production cross section by a factor of 1.9 for a muon collider and 3.4 for a electron–positron collider (Table 1). The ISR effect is asymmetric above and below the Higgs pole mass, and slightly shift the location of the peak cross section, as shown in Fig. 2.
- The total reduction factors for the on-resonance Higgs production cross section after convoluting the BES and ISR effects are 3.2 (7.1) for a muon collider with $R = 0.003\%$ ($R = 0.01\%$), and 11 (35) for a electron–positron collider with $R = 0.01\%$ ($R = 0.04\%$), as tabulated in the last column in Table 1.
- The background for the $h \to b\bar{b}$ channel is increased by a factor of seven due to the "radiative return" of the Z boson at lepton colliders and a cut on the minimal $b\bar{b}$ invariant mass of 100 GeV reduces such background, resulting in an increase of the tree-level estimate of the background by 20%. For a muon collider, both the $h \to b\bar{b}$ and $h \to WW^*$ contribute to the signal sensitivity. For an electron–positron collider, only the WW^* contributes due to the smallness of S/B for the $h \to b\bar{b}$ channel.
- For a muon collider resonant Higgs factory with our more robust study including both the BES and ISR effects, a 21 steps scan in the ±30 MeV window around the Higgs mass would provide percent level precision on the Higgs width and Higgs muon Yukawa coupling measurements, and sub MeV precision on the Higgs mass determination for various collider configurations, as tabulated in Table 3.
- For an electron–positron Higgs resonance factory, since a pre-scan to determine the precise Higgs mass is not feasible, one can achieve an 95% C.L. upper limit of 3 ∼ 5 (5 ∼ 10) times the SM electron Yukawa by scanning through the ±100 MeV window around Higgs mass with 100 (10) ab^{-1} integrated luminosity, as shown in Fig. 3.
- For an electron–positron Higgs resonance factory, increasing beam quality by reducing the beam energy spread does not increase the sensitivity to the Higgs electron Yukawa coupling, as *a priori* knowledge of the precise Higgs mass better than 10 MeV may not be available. Hence a first run at 240–250 GeV mode for an electron–positron collider maybe a step to make better result out of such resonance Higgs factory. A muon collider Higgs factory, on the other hand, is very complimentary and can provide sub MeV level of Higgs mass determination, which optimizes the sensitivity for the potential electron–positron Higgs resonance factory provided great beam quality can be achieved. In this case, a single run at fixed e^+e^- energy can achieve an upper limit of 1.7 (3) times the SM electron Yukawa for BES $R = 0.01\%$ with 100 (10) ab^{-1} integrated luminosity, as shown in Fig. 3.

Our study improves the existing analyses of the proposed future resonant Higgs factories and provides further guidance for the target accelerator designs with respect to the physical goals.

Acknowledgements

This work is supported in part by the U.S. Department of Energy under grant Nos. DE-FG02-95ER40896, DE-AC02-07CH11359 in part by PITT PACC. Fermilab is operated by Fermi Research Alliance, LLC under Contract No. DE-AC02-07CH11359 with the U.S. Department of Energy. Z.L. and T.H. thank the Kavli Institutes for Theoretical Physics at UC Santa Barbara and in China at the CAS, respectively, for their hospitality during the final stage of this paper.

Appendix A. Analytical formulas for the initial state radiation

We list a few commonly used analytical formulas for the ISR. We first introduce β_ℓ as a common loop factor for radiation effects

$$\beta_\ell = \frac{2\alpha}{\pi}\left(\log\frac{\hat{s}}{m_\ell^2} - 1\right), \tag{A.1}$$

where α is the fine-structure constant evaluated at the collision center of mass energy $\sqrt{\hat{s}}$ and m_ℓ is the charged lepton mass. The difference between the electron and muon is mainly carried in this factor. We present the ISR effect for various parameterization and expansion order in the following, where we annotate the dependence on charged leptons explicitly with the subscript ℓ.

- Kuraev–Fadin [25]:

$$f_{\ell\ell}^{\mathrm{ISR;KF}}(x;\hat{s}) = \int_x^1 dy\, 2 f_\ell^{\mathrm{KF}}(y;\hat{s}) f_\ell^{\mathrm{KF}}(\frac{x}{y};\hat{s}), \tag{A.2}$$

$$\text{with } f_\ell^{\mathrm{KF}}(x;\hat{s}) = \frac{\beta_\ell}{16}\left((8+3\beta_\ell)(1-x)^{\frac{\beta_\ell}{2}-1} - 4(1+x)\right).$$

- Nicrosini–Trentadue [26]:

$$\begin{aligned} f_{\ell\ell}^{\mathrm{ISR;\,NT}}(x;\hat{s}) &= \Delta_\ell \beta_\ell x^{\beta_\ell - 1} - \frac{1}{2}\beta_\ell(2-x) \\ &+ \frac{1}{8}\beta_\ell^2\left((2-x)(3\log(1-x) - 4\log x)\right. \\ &\left. - \frac{4\log(1-x)}{x} - 6 + x\right) + \mathcal{O}(\beta_\ell^3) \end{aligned} \tag{A.3}$$

with

$$\Delta_\ell = 1 + \frac{\alpha}{\pi}(\frac{3}{2}\log\frac{s}{m_\ell^2} - \frac{1}{3}\pi^2 - 2)$$
$$+ \frac{\alpha^2}{\pi^2}\left(\left(\frac{9}{8} - 2\zeta[2]\right)\log\frac{\hat{s}}{m_\ell^2}\right.$$
$$+\left(-\frac{45}{16} + \frac{11}{2}\zeta[2] + 3\zeta[3]\right)\log\frac{\hat{s}}{m_\ell^2} - \frac{6}{5}\zeta^2[2]$$
$$\left. - 6\zeta[2]\log 2 + \frac{3}{8}\zeta[2] + \frac{57}{12}\right),$$

where $\zeta[n]$ is the Euler–Riemann zeta function.

- Jadach–Ward–Was [27,18]:

$$f_{\ell\ell}^{\mathrm{ISR;\,JWW(a)}}(x;\hat{s}) = e^{\frac{\beta_\ell}{4} + \frac{\alpha}{\pi}\left(-\frac{1}{2} + \frac{\pi^2}{3}\right)} \frac{e^{-\gamma\beta_\ell}}{\Gamma[1+\beta_\ell]}\beta_\ell(1-x)^{\beta_\ell - 1}$$
$$f_{\ell\ell}^{\mathrm{ISR;\,JWW(b)}}(x;\hat{s}) = f_{\ell\ell}^{\mathrm{ISR;\,JWW(a)}}(x;\hat{s})\left(1 + \frac{\beta_\ell}{2} - \frac{1}{2}(1-x^2)\right)$$
$$f_{\ell\ell}^{\mathrm{ISR;\,JWW(c)}}(x;\hat{s}) \tag{A.4}$$
$$= f_{\ell\ell}^{\mathrm{ISR;\,JWW(a)}}(x;\hat{s})\left(1 + \frac{\beta_\ell}{2} + \frac{\beta_\ell^2}{8} - \frac{1}{2}(1-x^2)\right.$$
$$\left. + \beta_\ell\left(-\frac{1-x}{2} - \frac{1+3x^2}{8}\log x\right)\right),$$

where γ is the Euler–Mascheroni constant. Their numerical results are compared in Fig. 1 and the corresponding discussion in the text.

References

[1] ATLAS Collaboration, G. Aad, et al., Observation of a new particle in the search for the Standard Model Higgs boson with the ATLAS detector at the LHC, Phys. Lett. B 716 (2012) 1–29, arXiv:1207.7214.

[2] CMS Collaboration, S. Chatrchyan, et al., Observation of a new boson at a mass of 125 GeV with the CMS experiment at the LHC, Phys. Lett. B 716 (2012) 30–61, arXiv:1207.7235.

[3] V.D. Barger, M.S. Berger, J.F. Gunion, T. Han, s Channel Higgs boson production at a muon muon collider, Phys. Rev. Lett. 75 (1995) 1462–1465, arXiv:hep-ph/9504330.

[4] V.D. Barger, M.S. Berger, J.F. Gunion, T. Han, Higgs Boson physics in the s channel at $\mu^+\mu^-$ colliders, Phys. Rep. 286 (1997) 1–51, arXiv:hep-ph/9602415.

[5] T. Han, Z. Liu, Potential precision of a direct measurement of the Higgs boson total width at a muon collider, Phys. Rev. D 87 (3) (2013) 033007, arXiv:1210.7803.

[6] Y. Alexahin, et al., The case for a muon collider Higgs factory, in: Community Summer Study 2013: Snowmass on the Mississippi, CSS2013, Minneapolis, MN, USA, July 29–August 6, 2013, 2013, arXiv:1307.6129.

[7] Y. Alexahin, et al., Muon collider Higgs factory for snowmass 2013, in: Community Summer Study 2013: Snowmass on the Mississippi, CSS2013, Minneapolis, MN, USA, July 29–August 6, 2013, 2013, arXiv:1308.2143.

[8] C. Rubbia, A complete demonstrator of a muon cooled Higgs factory, arXiv:1308.6612.

[9] H. Baer, T. Barklow, K. Fujii, Y. Gao, A. Hoang, S. Kanemura, J. List, H.E. Logan, A. Nomerotski, M. Perelstein, et al., The international linear collider technical design report – volume 2: physics, arXiv:1306.6352.

[10] TLEP Design Study Working Group Collaboration, M. Bicer, et al., First look at the physics case of TLEP, J. High Energy Phys. 01 (2014) 164, arXiv:1308.6176.

[11] M. Ahmad, et al., CEPC-SPPC preliminary conceptual design report, volume I: physics and detector, http://cepc.ihep.ac.cn/preCDR/volume.html, 2015.

[12] A. Apyan, et al., CEPC-SPPC preliminary conceptual design report, volume II: accelerator, http://cepc.ihep.ac.cn/preCDR/volume.html, 2015.

[13] S. Dawson, et al., Working group report: Higgs boson, in: Community Summer Study 2013: Snowmass on the Mississippi, CSS2013, Minneapolis, MN, USA, July 29–August 6, 2013, 2013, arXiv:1310.8361.

[14] D. d'Enterria, G. Wojcik, R. Aleksan, Search of resonant s-channel Higgs production at FCC-ee, https://dde.web.cern.ch/dde/presentations/fcc_phys_cern_jun14/fcc_Higgs_schannel_jun14.pdf.

[15] W. Altmannshofer, J. Brod, M. Schmaltz, Experimental constraints on the coupling of the Higgs boson to electrons, J. High Energy Phys. 05 (2015) 125, arXiv:1503.04830.

[16] S. Dittmaier, et al., Handbook of LHC Higgs cross sections: 2. differential distributions, arXiv:1201.3084.

[17] M. Greco, On the study of the Higgs properties at a muon collider, Mod. Phys. Lett. A 30 (39) (2015) 1530031, arXiv:1503.05046.

[18] S. Jadach, R.A. Kycia, Lineshape of the Higgs boson in future lepton colliders, Phys. Lett. B 755 (2016) 58–63, arXiv:1509.02406.

[19] D.R. Yennie, S.C. Frautschi, H. Suura, The infrared divergence phenomena and high-energy processes, Ann. Phys. 13 (1961) 379–452.

[20] SLAC-SP-017 Collaboration, J.E. Augustin, et al., Discovery of a narrow resonance in e^-e^- annihilation, Phys. Rev. Lett. 33 (1974) 1406–1408, Adv. Exp. Phys. 5 (1976) 141, ADONE, Frascati also observed a couple days after SLAC and published in PRL just after the SLAC paper: C. Bacci, et al., Phys. Rev. Lett. 33 (1974) 1408.

[21] M. Greco, G. Pancheri-Srivastava, Y. Srivastava, Radiative corrections for colliding beam resonances, Nucl. Phys. B 101 (1975) 234.

[22] M. Greco, G. Pancheri-Srivastava, Y. Srivastava, Radiative corrections to $e^+e^- \to \mu^+\mu^-$ around the Z_0, Nucl. Phys. B 171 (1980) 118, Nucl. Phys. B 197 (1982) 543 (Erratum).

[23] J.R. Ellis, R. Peccei, Physics at LEP 1, Tech. Rep. CERN-86-02-V-1, CERN-YELLOW-86-02-V-1, CERN-86-02, 1986.

[24] J.R. Ellis, R. Peccei, Physics at LEP 2, Tech. Rep. CERN-86-02-V-2, CERN-YELLOW-86-02-V-2, CERN-86-02, 1986.

[25] E.A. Kuraev, V.S. Fadin, On radiative corrections to e^+e^- single photon annihilation at high-energy, Sov. J. Nucl. Phys. 41 (1985) 466–472, Yad. Fiz. 41 (1985) 733.

[26] O. Nicrosini, L. Trentadue, Soft photons and second order radiative corrections to $e^+e^- \longrightarrow Z^0$, Phys. Lett. B 196 (1987) 551.

[27] S. Jadach, B.F.L. Ward, Z. Was, Coherent exclusive exponentiation for precision Monte Carlo calculations, Phys. Rev. D 63 (2001) 113009, arXiv:hep-ph/0006359.

[28] N. Chakrabarty, T. Han, Z. Liu, B. Mukhopadhyaya, Radiative return for heavy Higgs boson at a muon collider, Phys. Rev. D 91 (1) (2015) 015008, arXiv:1408.5912.

[29] M. Karliner, M. Low, J.L. Rosner, L.-T. Wang, Radiative return capabilities of a high-energy, high-luminosity e^+e^- collider, Phys. Rev. D 92 (3) (2015) 035010, arXiv:1503.07209.

[30] A. Conway, H. Wenzel, Higgs measurements at a muon collider, arXiv:1304.5270.

[31] FCC-ee machine parameters, http://tlep.web.cern.ch/content/machine-parameters.

Volume 77B, number 3 PHYSICS LETTERS 14 August 1978

COHERENT STATE APPROACH TO THE INFRA-RED BEHAVIOUR OF NON-ABELIAN GAUGE THEORIES

M. GRECO, F. PALUMBO, G. PANCHERI-SRIVASTAVA [1] and Y. SRIVASTAVA [1]
INFN, Laboratori Nazionali di Frascati, Rome, Italy

Received 31 January 1978

The infrared behaviour of matrix elements between coherent states of definite color is studied in non-abelian gauge theories. The matrix elements are shown to be finite to the lowest non-trivial order, and to all orders if the soft meson formula for real gluons holds to all orders. Factorization occurs in the fixed angle regime.

In this letter we study the infrared (IR) behaviour of non-abelian gauge theories (NAGT) using the coherent state formalism. This method has been applied [1] in QED providing one with a definition of matrix elements which are (i) free from IR divergences, (ii) factorizable and directly comparable with the experimental cross sections. This approach is exactly equivalent to the standard one of summing cross sections for real and virtual photon emission [1,2].

Our extension to NAGT is based on the (assumed) validity, to all orders, of the so-called soft meson formula [3]. This formula has been proved to all orders in the leading logarithms for virtual gluons [4].

It has also been checked for single-gluon emission in different processes [5], and for double-gluon emission both in quark scattering in an external colorless potential [6] and in quark–quark scattering [7].

In all cases the leading divergences cancel out in the inclusive cross sections after color average over both initial and final states, in agreement with the Lee and Nauenberg theorem [8]. Due to the color average, however, nothing can be said about cross sections for states of definite color. The coherent state formalism, on the other hand, is such as to provide one with matrix elements which are free from IR divergences for any initial or final color states. This formalism, therefore, turns out to be not equivalent to the standard one of summing cross sections for real and virtual gluon emission, in contrast to QED.

The factorization properties of the new matrix elements are also different from QED. In fact, factorization occurs only in the fixed angle regime (see below for definitions), as also found for the inclusive color averaged cross section [5–7].

We shall not discuss in this letter the connection between our results and physically measurable quantities.

For given initial and final states $|i\rangle$ and $|f\rangle$, we define the corresponding coherent states $|\tilde{i}\rangle$ and $|\tilde{f}\rangle$:

$$|\tilde{i}\rangle = e^{i\Lambda_i}|i\rangle, \quad |\tilde{f}\rangle = e^{i\Lambda_f}|f\rangle \,, \tag{1}$$

with

$$\Lambda = \frac{1}{(2\pi)^4} \int d^4k \, j_\mu^c(k) A_\mu^c(-k). \tag{2}$$

In eq. (2) $A_\mu^c(x)$ is the quantized gluon field of color index c, and $j_\mu^c(k)$ is the "classical" current

[1] Physics Department, Northeastern University, Boston, MA, USA. Work supported in part by a grant from the National Science Foundation, USA.

$$j_\mu^c(k) = \frac{\mathrm{i}}{(2\pi)^{3/2}} \bar{g}(k) \sum_\alpha \frac{\eta_\alpha p_{\alpha\mu}}{(p_\alpha k)} t^c(\alpha) \equiv \sum_\alpha j_{\alpha\mu}(k) t^c(\alpha), \tag{3}$$

where α is the label of particles in initial and final states, $\eta_\alpha = +1$ for incoming particles and outgoing antiparticles and $\eta_\alpha = -1$ otherwise. The operators $t^c(\alpha)$ are the appropriate generators of the color group for the αth particle ($t^c(\alpha)$ must be replaced by $t^{c*}(\alpha)$ for outgoing particles) and $\bar{g}(k)$ is the effective coupling constant for pure Yang–Mills theory, which appears in the soft meson formula for virtual gluons [4] (see discussion later).

Let us define new S-matrix elements:

$$\bar{M} = \langle \tilde{\mathrm{f}} | S | \tilde{\mathrm{i}} \rangle = \langle \mathrm{f} | \mathrm{e}^{-\mathrm{i}\Lambda_\mathrm{f}} S\, \mathrm{e}^{\mathrm{i}\Lambda_\mathrm{i}} | \mathrm{i} \rangle \equiv \langle \mathrm{f} | \bar{S} | \mathrm{i} \rangle. \tag{4}$$

In terms of a perturbative expansion in $\bar{g}(k)$,

$$\bar{M} = \sum_{n=0}^{\infty} \bar{M}_n, \tag{5}$$

with

$$\bar{M}_n = \sum_{k=0}^{n} \sum_{l=0}^{k} \langle \mathrm{f} | \frac{(-\mathrm{i}\Lambda_\mathrm{f})^{k-l}}{(k-l)!} S_{n_0,l} \frac{(\mathrm{i}\Lambda_\mathrm{i})^{n-k}}{(n-k)!} | \mathrm{i} \rangle, \tag{6}$$

where n_0 corresponds to the lowest-order non-vanishing transition amplitude. $\bar{M}_n$ and $S_{n_0,l}$ are of order n_0 in the bare coupling constant but are of order n and l, respectively, in the effective coupling constant $\bar{g}(k)$. This follows from eq. (3) and the results of ref. [4].

Let us concentrate on $\bar{M}_2$, the first radiatively corrected matrix element. One of the three terms in $\bar{M}_2$ is given by

$$\langle \mathrm{f} | S_{n_0,1}(\mathrm{i}\Lambda_\mathrm{i}) | \mathrm{i} \rangle = \frac{\mathrm{i}}{(2\pi)^4} \int \mathrm{d}^4k \sum_{\substack{\alpha \in \{\mathrm{i}\} \\ \sigma}} j_{\alpha\mu}(k) \langle \mathrm{f} | S_{n_0,1} A_\mu^c(-k) | \mathrm{i}, \lambda_\alpha \to \sigma \rangle \langle \sigma | t^c | \lambda_\alpha \rangle, \tag{7}$$

where λ_α is the color index of the αth particle in the initial state i.

The matrix element appearing in eq. (7) is evaluated by using the assumed soft meson formula for the emission of a soft gauge vector from any on-shell process:

$$\langle \mathrm{f} | T_\nu^a(k) | \mathrm{i} \rangle = \sum_{\substack{\gamma \in \{\mathrm{i}\} \\ \lambda}} \bar{g}(k) \eta_\gamma \frac{p_{\gamma\nu}}{(p_\gamma \cdot k)} \langle \lambda | t^a | \lambda_\gamma \rangle \langle \mathrm{f} | S | \mathrm{i}, \lambda_\gamma \to \lambda \rangle + \sum_{\gamma \in \{\mathrm{f}\}} \bar{g}(k) \eta_\gamma \frac{p_{\gamma\nu}}{(p_\gamma \cdot k)} \langle \lambda_\gamma | t^\alpha | \lambda \rangle \langle \mathrm{f}, \lambda_\gamma \to \lambda | S | \mathrm{i} \rangle. \tag{8}$$

Then eq. (7) becomes

$$\begin{aligned} \langle \mathrm{f} | S_{n_0,1}(\mathrm{i}\Lambda_\mathrm{i}) | \mathrm{i} \rangle = \int \mathrm{d}^4k\, \delta(k^2) \theta(k_0) \Bigg\{ \sum_{\alpha \in \{\mathrm{i}\},\sigma} j_{\alpha\mu}(k) \langle \sigma | t^c | \lambda_\alpha \rangle \Bigg\} \\ \times \Bigg\{ \sum_{\alpha' \in \{\mathrm{i}\},\sigma'} j_{\alpha'}^\mu(-k) \langle \sigma' | t^c | \lambda_{\alpha'}(\sigma) \rangle \langle \mathrm{f} | S_{n_0,0} | \mathrm{i}, \begin{bmatrix} \lambda_\alpha \to \sigma \\ \lambda_{\alpha'}(\sigma) \to \sigma' \end{bmatrix} \rangle \\ + \sum_{\beta' \in \{\mathrm{f}\},\rho'} j_{\beta'}^\mu(-k) \langle \lambda_{\beta'} | t^c | \rho' \rangle \langle \mathrm{f}, \lambda_{\beta'} \to \rho' | S_{n_0,0} | \mathrm{i}, \lambda_\alpha \to \sigma \rangle \Bigg\}, \end{aligned} \tag{9}$$

where in the state $|\lambda_{\alpha'}(\sigma)\rangle$ the color index is $\lambda_{\alpha'}$ for $\alpha' \neq \alpha$ and σ otherwise. A similar evaluation of the other

Volume 77B, number 3 PHYSICS LETTERS 14 August 1978

terms in $\bar{M}_2$ leads to the result:

$$\bar{M}_2 = \langle \mathrm{f}|S_{n_0,2}|\mathrm{i}\rangle + \langle \mathrm{f}|R_{n_0,2}|\mathrm{i}\rangle, \tag{10}$$

with

$$\begin{aligned}\langle \mathrm{f}|R_{n_0,2}|\mathrm{i}\rangle = \tfrac{1}{2}\int \mathrm{d}^4k\,\delta(k^2)\theta(k_0)\Bigg\{ &\sum_{\alpha,\alpha'\in\{\mathrm{i}\},\sigma,\sigma'} j_{\alpha\mu}(k) j^{\mu}_{\alpha'}(-k)\langle\sigma|t^c|\lambda_\alpha\rangle\langle\sigma'|t^c|\lambda_{\alpha'}(\sigma)\rangle\langle \mathrm{f}|S_{n_0,0}|\mathrm{i},\begin{bmatrix}\lambda_\alpha\to\sigma\\ \lambda_{\alpha'}(\sigma)\to\sigma'\end{bmatrix}\rangle \\ &+\sum_{\alpha\in\{\mathrm{i}\},\beta'\in\{\mathrm{f}\},\sigma,\rho'} j_{\alpha\mu}(k) j^{\mu}_{\beta'}(-k)\langle\sigma|t^c|\lambda_\alpha\rangle\langle\lambda_{\beta'}|t^c|\rho'\rangle\langle \mathrm{f},\lambda_{\beta'}\to\rho'|S_{n_0,0}|\mathrm{i},\lambda_\alpha\to\sigma\rangle \\ &+\sum_{\beta\in\{\mathrm{f}\},\alpha'\in\{\mathrm{i}\},\rho,\sigma'} j_{\beta\mu}(k) j^{\mu}_{\alpha'}(-k)\langle\sigma'|t^c|\lambda_{\alpha'}\rangle\langle\lambda_\beta|t^c|\rho\rangle\langle \mathrm{f},\lambda_\beta\to\rho|S_{n_0,0}|\mathrm{i},\lambda_{\alpha'}\to\sigma'\rangle \\ &+\sum_{\beta,\beta'\in\{\mathrm{f}\},\rho,\rho'} j_{\beta\mu}(k) j^{\mu}_{\beta'}(-k)\langle\lambda_\beta|t^c|\rho\rangle\langle\lambda_{\beta'}(\rho)|t^c|\rho'\rangle\langle \mathrm{f},\begin{bmatrix}\lambda_\beta\to\rho\\ \lambda_{\beta'}(\rho)\to\rho'\end{bmatrix}|S_{n_0,0}|\mathrm{i}\rangle\ .\end{aligned} \tag{11}$$

In shorthand notation

$$\langle \mathrm{f}|R_{n_0,2}|\mathrm{i}\rangle = \tfrac{1}{2}\int \mathrm{d}^4k\,\delta(k^2)\theta(k_0)\langle \mathrm{f}|\{S_{n_0,0}\, j^c_{\mathrm{in}\,\mu}(k) j^{c\mu}_{\mathrm{in}}(-k) + 2 j^c_{\mathrm{out}\,\mu}(k) S_{n_0,0}\, j^{c\mu}_{\mathrm{in}}(-k) + j^c_{\mathrm{out}\,\mu}(k) j^{c\mu}_{\mathrm{out}}(-k) S_{n_0,0}\}|\mathrm{i}\rangle. \tag{12}$$

On the other hand, for the pure virtual gluon contribution $\langle \mathrm{f}|S_{n_0,2}|\mathrm{i}\rangle$ one gets the same result, but for the replacement of $\delta(k^2)\theta(k_0)$ by $(-\mathrm{i}/2\pi)(1/k^2)$.

One concludes therefore that $\bar{M}_2$ is finite and has the following form:

$$\bar{M}_2 = \tfrac{1}{2}\int \mathrm{d}^4k\left[\delta(k^2)\theta(k_0) - \frac{\mathrm{i}}{2\pi}\frac{1}{k^2}\right]\langle \mathrm{f}|\{ j^c_{\mathrm{out}\,\mu}(k) j^{c\mu}_{\mathrm{out}}(-k)\bar{S}_0 + 2 j^c_{\mathrm{out}\,\mu}(k)\bar{S}_0\, j^{c\mu}_{\mathrm{in}}(-k) + \bar{S}_0\, j^c_{\mathrm{in}\,\mu}(k) j^{c\mu}_{\mathrm{in}}(-k)\}|\mathrm{i}\rangle. \tag{13}$$

By finite we mean that the standard infrared singularities have been eliminated. It is always possible however that the complete $\bar{g}(k)$ is such that $\bar{M}$ is still not convergent.

We are now going to relate the matrix element $R_{n_0,2}$ to the probability of emission of one real gluon. From eq. (8), rewritten in short as

$$\langle \mathrm{f}|T^c_\mu(k)|\mathrm{i}\rangle = \langle \mathrm{f}|j^c_{\mathrm{out}\,\mu}(k) S_{n_0,0}|\mathrm{i}\rangle + \langle \mathrm{f}|S_{n_0,0}\, j^c_{\mathrm{in}\,\mu}(k)|\mathrm{i}\rangle, \tag{14}$$

we have

$$\begin{aligned}&\int \mathrm{d}^4k\,\delta(k^2)\theta(k_0)\langle \mathrm{f}|T^c_\mu(k)|\mathrm{i}\rangle\langle \mathrm{f}|T^{c\mu}(k)|\mathrm{i}\rangle^* \\ &\quad= \int \mathrm{d}^4k\,\delta(k^2)\theta(k_0)\{\langle \mathrm{f}|j^c_{\mathrm{out}}(k) S_{n_0,0}|\mathrm{i}\rangle\langle \mathrm{i}|S^+_{n_0,0}\, j^c_{\mathrm{out}}(-k)|\mathrm{f}\rangle + \langle \mathrm{f}|j^c_{\mathrm{out}}(k) S_{n_0,0}|\mathrm{i}\rangle\langle \mathrm{i}|j^c_{\mathrm{in}}(-k) S^+_{n_0,0}|\mathrm{f}\rangle \\ &\quad+ \langle \mathrm{f}|S_{n_0,0}\, j^c_{\mathrm{in}}(k)|\mathrm{i}\rangle\langle \mathrm{i}|S^+_{n_0,0}\, j^c_{\mathrm{out}}(-k)|\mathrm{f}\rangle + \langle \mathrm{f}|S_{n_0,0}\, j^c_{\mathrm{in}}(k)|\mathrm{i}\rangle\langle \mathrm{i}|j^c_{\mathrm{in}}(-k) S^+_{n_0,0}|\mathrm{f}\rangle\}.\end{aligned} \tag{15}$$

On the other hand, from eq. (5) we have

$$\begin{aligned}|\bar{M}|^2 &= |\bar{M}_0 + \bar{M}_2 + \ldots|^2 = |\langle \mathrm{f}|S_{n_0,0}|\mathrm{i}\rangle|^2 \\ &+ \{\langle \mathrm{f}|S_{n_0,0}|\mathrm{i}\rangle\langle \mathrm{f}|S_{n_0,2}|\mathrm{i}\rangle^* + \langle \mathrm{f}|S_{n_0,0}|\mathrm{i}\rangle\langle \mathrm{f}|R_{n_0,2}|\mathrm{i}\rangle^* + \mathrm{c.c.}\} + \text{higher orders}.\end{aligned} \tag{16}$$

From eqs. (12) and (16) it is clear that the term $R_{n_0,2}$ in $\bar{M}$ is equivalent to the single gluon bremsstrahlung only upon average over color of the external particles.

A recursion formula is conjectured for the matrix element $\bar{M}$,

$$\langle \mathrm{f}|\bar{S}|\mathrm{i}\rangle_{n+2} = \tfrac{1}{2} \int \mathrm{d}^4k \left[\delta(k^2)\theta(k^0) - \frac{\mathrm{i}}{2\pi}\frac{1}{k^2}\right] \langle \mathrm{f}|\{ j^{c}_{\mathrm{out}\,\mu}(k) j^{c\mu}_{\mathrm{out}}(-k)\bar{S}_n + 2 j^{c}_{\mathrm{out}\,\mu}(k)\bar{S}_n j^{c\mu}_{\mathrm{in}}(-k) + \bar{S}_n j^{c}_{\mathrm{in}\,\mu}(k) j^{c\mu}_{\mathrm{in}}(-k)\}|\mathrm{i}\rangle, \tag{17}$$

which is based on the extension of the soft-meson formula for virtual gluon to $\bar{S}$. Eq. (13) is the first step towards eq. (17), which proves the finiteness of $\bar{M}$. As for the lowest-order calculation, the equivalence of the present calculation to the nth order inclusive cross section holds only after average over initial and final color states, in contrast to the abelian case.

The virtual gluon part of eq. (17) has been derived in refs. [3] and [4] in the form of a differential equation.

In the fixed angle regime, when all the invariant squared energies and momentum transfers become large, the r.h.s. of eq. (17) gets factorized and $\bar{M}$ exponentiates as in the abelian case. One finds

$$\langle \mathrm{f}|\bar{S}|\mathrm{i}\rangle_{n+2} = \tfrac{1}{2} B \Big(\sum_{\alpha} C_\alpha\Big) \langle \mathrm{f}|\bar{S}|\mathrm{i}\rangle_n , \tag{18}$$

where C_α is the eigenvalue of the Casimir operator for the αth particle, the sum runs over both the initial and final particles, and

$$B = B_{\alpha\beta} = \int \mathrm{d}^4k \left[\delta(k^2)\theta(k^0) - \frac{\mathrm{i}}{2\pi}\frac{1}{k^2}\right] \frac{(p_\alpha p_\beta)}{(p_\alpha k)(p_\beta k)} \frac{\bar{g}^2(k)}{(2\pi)^3} . \tag{19}$$

Notice that in the same regime, exponentiation occurs for the real bremsstrahlung only for the color averaged cross sections. Details of this result shall be presented elsewhere.

Finally, we compare our formalism with some perturbative calculations, which in particular allows us to justify the introduction of $\bar{g}(k)$ in eq. (3) ‡1. More precisely, let us consider massive quark scattering in an external colourless potential. Then eqs. (17)–(19) give

$$\langle \mathrm{f}|\bar{S}|\mathrm{i}\rangle_{n+2} = \left\{ \tfrac{1}{2} C_{\mathrm{F}} M(q^2) \int_0^{\ln\Delta} \mathrm{d}(\ln k)\, \bar{g}^2(k) \right\} \langle \mathrm{f}|\bar{S}|\mathrm{i}\rangle_n , \tag{20}$$

where C_{F} is the eigenvalue of the Casimir operator for the fundamental representation, Δ is the fractional energy loss and $M(q^2) = (1/2\pi^2)\ln(q^2/m^2)$ $(q^2 \gg m^2)$. This leads to

$$\mathrm{d}\sigma \propto |\langle \mathrm{f}|\bar{S}|\mathrm{i}\rangle|^2 = \exp\left\{ C_{\mathrm{F}} M(q^2) \int_0^{\ln\Delta} \mathrm{d}(\ln k)\bar{g}^2(k) \right\} |\langle \mathrm{f}|\bar{S}|\mathrm{i}\rangle_0|^2 . \tag{21}$$

Comparison with the calculation of the inclusive cross section by Frenkel et al. [6] then gives, for the leading terms,

$$\bar{g}^2(t) = g^2\left(1 - \frac{11}{24\pi^2} C_{\mathrm{YM}} g^2 t + \dots\right) = g^2\left(1 + \frac{11}{24\pi^2} C_{\mathrm{YM}} g^2 t\right)^{-1} , \tag{22}$$

where C_{YM} is the eigenvalue of the Casimir operator for the adjoint representation and $t \sim \ln k$.

Summarizing, we have constructed a coherent state formalism for NAGT, showing that matrix elements between coherent states of definite color are finite and factorized in the fixed angle regime.

Two of us (M.G. and F.P.) are grateful to A. Ukawa for very useful discussions.

‡1 In this context see also ref. [9].

Volume 77B, number 3 PHYSICS LETTERS 14 August 1978

References

[1] V. Chung, Phys. Rev. B140 (1965) 1110;
M. Greco and G. Rossi, Nuovo Cimento 50 (1967) 168.
[2] T. Kibble, J. Math. Phys. 9 (1968) 315; Phys. Rev. 173 (1968) 1527; 174 (1968) 1882; 175 (1968) 1624;
P. Kulish and L. Faddeev, Teor. Math. Fiz. 4 (1970) 153; Engl. transl. Theor. Math. Phys. (USSR) 4 (1970) 745;
J. Storrow, Nuovo Cimento 54A (1968) 15;
S. Schweber, in: Cargese Summer School Lectures, eds. M. Levy and J. Sucher (Gordon and Breach, New York, 1972);
D. Zwanziger, Phys. Rev. D11 (1975) 3481, 3504;
M. Greco, G. Pancheri-Srivastava and Y. Srivastava, Nucl. Phys. B101 (1975) 234.
[3] J.M. Cornwall and G. Tiktopoulos, Phys. Rev. D13 (1976) 3370; D15 (1977) 2937.
[4] T. Kinoshita and A. Ukawa, Phys. Rev. D15 (1977) 1596; D16 (1977) 332.
[5] Y.P. Yao, Phys. Rev. Lett. 36 (1976) 653;
T. Appelquist, J. Carazzone, H. Kluberg-Stern and M. Roth, Phys. Rev. Lett. 36 (1976) 768;
L. Tyburski, Phys. Rev. Lett. 37 (1976) 319.
[6] J. Frenkel, R. Meuldermans, I. Mohammad and J.C. Taylor, Phys. Lett. 64B (1976) 211; Nucl. Phys. B121 (1977) 58.
[7] M. Greco, F. Palumbo, G. Pancheri-Srivastava and Y. Srivastava, to be published.
[8] T.D. Lee and N. Nauenberg, Phys. Rev. B133 (1964) 1549.
[9] E. Poggio, Phys. Rev. Lett. 36 (1976) 1511.

Volume 79B, number 4,5 PHYSICS LETTERS 4 December 1978

MASS SINGULARITIES AND COHERENT STATES IN GAUGE THEORIES

G. CURCI
CERN, Geneva, Switzerland

and

M. GRECO
Laboratori Nazionali dell'INFN, Frascati, Italy

Received 18 July 1978

The problem of mass singularities is studied using the formalism of coherent states. The introduction of new states which account for both soft and hard radiation effects provides one with matrix elements which are free from infra-red and mass singularities at the leading logarithmic approximation. In QED our results coincide with those obtained from the renormalization group equations. The generalization to QCD agrees with perturbative calculations and provides a simple framework for evaluating the finite corrections to hard processes.

The problem of infra-red singularities in QED has been extensively studied in the literature, using mainly two different approaches. In the first and more conventional one [1] the infra-red structure of the on-shell form factors is inspected in perturbation theory, where various proofs have been given of the well-known exponentiation of the infra-red singularities. As is well known, in the calculation of inclusive cross sections for physical processes these divergences are exactly cancelled at all orders by those arising from real emission, in agreement with the Bloch–Nordsieck theorem [2], generalized by Kinoshita and Lee and Nauenberg [3].

In the second approach [4] the finiteness of observable cross sections has been the motivation for defining a finite S matrix, using the formalism of the coherent states. This is much closer to the physical reality because of the introduction in the theory of new states which are degenerate in the number of soft photons. The Kinoshita–Lee–Nauenberg theorem is then automatically satisfied and the new matrix elements are therefore finite.

The infra-red structure of non-abelian gauge theories has been investigated much the same as in QED. Using the first type of approach [5] and at the approximation where only leading powers of infra-red logarithms are kept, the on-shell form factors have been shown to satisfy the same infra-red differential equation found in QED, the only modification being the appearance of the effective coupling constant $\bar{g}^2(\tau)$ for the pure Yang–Mills theory instead of g^2 [6]. Furthermore explicit calculations [5] of observable quantities as inclusive cross sections, have also been found to be finite, in agreement with the Kinoshita–Lee–Nauenberg theorem.

These results have motivated the extension to non-abelian gauge theories of the coherent state formalism [7] which indeed provides one with matrix elements free from infra-red divergence for definite colour states, at the same leading logarithm approximations. This has been achieved by extending the concept of classical currents associated to the external particles to incorporate both the new properties of colour and the appearance of the effective coupling constant $\bar{g}^2(\tau)$.

In the present letter we further extend the formalism of coherent states to the problem of mass singularities in QED. The obvious motivation for this analysis is the very close connection between hard processes and mass singularities in quantum chromodynamics (QCD), as recently emphasized by various authors [8].

To this aim we enlarge the concept of soft photons by considering in addition to the usual low-energy photons ($k \leqslant \Delta\omega < E$), also the hard photons ($\Delta\omega \leqslant k < E$) which are emitted with an angle less than or equal to an arbitrarily small but fixed value δ. In other words we include in the formalism all photons which have either small energies or high energies but small transverse momenta ($k_\perp^2 \leqslant E^2\delta^2$).

Such an extension provides one with new matrix elements free of mass singularities at all orders of perturbation theory, in the leading logarithm approximation, in agreement with the Kinoshita–Lee–Nauenberg theorem. As a nice feature of the formalism the contribution of hard and collinear photons from any external leg of a diagram is easily obtained in exponentiated form. Our results are then simply generalized to QCD by virtue of the formalism developed in ref. [7] and provide a very simple tool for evaluating to all orders the finite corrections to hard processes.

To be more specific, let us consider electron scattering in an external field. After cancellation of the infrared divergences from real and virtual soft photons ($k \leqslant \Delta\omega$) the inclusive cross section, at the lowest radiative correction, is given by ($-q^2 \gg m^2$)

$$\sigma_{\rm incl} = \sigma_0 \left\{ 1 + 2\frac{\alpha}{\pi}\left[\left(\ln\frac{-q^2}{m^2} - 1\right)\ln\frac{\Delta\omega}{E} + \frac{3}{4}\ln\frac{-q^2}{m^2} + O(1)\right]\right\}, \tag{1}$$

where $\alpha = e^2/4\pi$ is the usual on-shell charge. The contribution of vacuum polarization is not included in eq. (1) and will be discussed later with the renormalization problem. As $m \to 0$ the mass singularities of eq. (1) are exactly cancelled by adding the contributions of the states which are degenerate with the initial and final electron. More explicitly, the total hard cross section

$$\sigma_{\rm hard} = \sigma[e(p-k) + \gamma(k) \to e(p')] + \sigma[e(p) \to e(p'-k') + \gamma(k')],$$

is given by

$$\sigma_{\rm hard} = \sigma_0 \left\{ 2\frac{\alpha}{\pi}\left[\left(\ln\frac{E^2\delta^2}{m^2} - 1\right)\left(\ln\frac{E}{\Delta\omega} - \frac{3}{4}\right) + O(1)\right]\right\}, \tag{2}$$

having limited the hard photon phase space as $\Delta\omega < k$, $k' < E$ and $0 < \theta_k, \theta_{k'} < \delta$. There, as is well known, the superinclusive cross section contains no mass singularities:

$$\sigma_{\rm super} \equiv \sigma_{\rm incl} + \sigma_{\rm hard} = \sigma_0 \left\{ 1 + 2\frac{\alpha}{\pi}\left[\ln\frac{-q^2}{k_{\perp\,\rm max}^2}\cdot\ln\frac{\Delta\omega}{E} + \frac{3}{4}\ln\frac{-q^2}{k_{\perp\,\rm max}^2} + O(1)\right]\right\}, \tag{3}$$

with $k_{\perp\,\rm max}^2 = E^2\delta^2$.

To improve eqs. (1) and (2) to all orders, to maximally logarithmic terms, we use the technique of the renormalization group [9] for the inclusive cross section [10] and then obtain the hard component by the Kinoshita–Lee–Nauenberg theorem. We find

$$\sigma_{\rm incl} = \sigma_0 \exp\left(-2\frac{\alpha}{\pi}\ln\frac{\Delta\omega}{E}\right)\exp\left[\int_{\alpha/\pi}^{(\alpha/\pi)_{\rm incl}} \frac{dx}{x}\frac{\gamma(x)}{\beta(x)}\right], \tag{4}$$

and

$$\sigma_{\rm hard} = \sigma_0 \exp\left(2\frac{\alpha}{\pi}\ln\frac{\Delta\omega}{E}\right)\exp\left\{-\int_{\alpha/\pi}^{(\alpha/\pi)_{\rm hard}} \frac{dx}{x}\frac{\gamma(x)}{\beta(x)}\right\}, \tag{5}$$

where $\gamma(x) = x(4\ln(\Delta\omega/E) + 3)$ and $\beta(x) = x\beta_1$ ($\beta_1 = \frac{2}{3}$). In the above equations $(\alpha/\pi)_{\rm incl}$ and $(\alpha/\pi)_{\rm hard}$ are the running coupling constants for the inclusive and hard problem, namely the solution of the equations

$$\left\{\frac{\partial}{\partial t_{\substack{\rm incl\\ \rm hard}}} - \beta\left(\frac{\alpha}{\pi}\right)\frac{\alpha}{\pi}\frac{\partial}{\partial\alpha/\pi}\right\}\left(\frac{\alpha}{\pi}\right)_{\substack{\rm incl\\ \rm hard}} = 0, \tag{6}$$

where $t_{\rm incl} = \frac{1}{2}\ln(-q^2/m^2)$ and $t_{\rm hard} = \frac{1}{2}\ln(E^2\delta^2/m^2)$. We therefore obtain, for the superinclusive cross section

$$\sigma_{\rm super} = \sigma_0 \exp\left\{-\int_{(\alpha/\pi)_{\rm incl}}^{(\alpha/\pi)_{\rm hard}} \frac{dx}{x}\frac{\gamma(x)}{\beta(x)}\right\}. \tag{7}$$

To avoid mass singularities associated with on-shell charge renormalization we use as an expansion parameter $(\alpha/\pi)_{\rm incl}$, which physically corresponds to renormalize the charge at the value q^2. Consistently we have omitted the vacuum polarization contributions.

Volume 79B, number 4,5 PHYSICS LETTERS 4 December 1978

Finally, expressing $(\alpha/\pi)_{\rm hard}$ in terms of $(\alpha/\pi)_{\rm incl} \equiv \bar{\alpha}/\pi$, we obtain

$$\sigma_{\rm super} = \sigma_0 \exp\left\{\frac{4}{\beta_1}\left(\ln\frac{\Delta\omega}{E} + \frac{3}{4}\right)\ln\left(1 + \frac{\bar{\alpha}}{2\pi}\beta_1 \ln\frac{-q^2}{E^2\delta^2}\right)\right\}. \tag{8}$$

We are now going to discuss the same problem from the approach of the coherent states. As is well known [4], for given initial and final states $|i\rangle$ and $|f\rangle$, defining the corresponding coherent states $|\tilde{i}\rangle$ and $|\tilde{f}\rangle$ in terms of the classical currents associated to the external particles as

$$|\tilde{i}, f\rangle = e^{i\Lambda_{i,f}}|i, f\rangle, \tag{9}$$

with

$$\Lambda_{i,f} = \frac{1}{(2\pi)^4}\int d^4k\, j_\mu^{(i,f)}(k) A_\mu(-k), \tag{10}$$

the new S-matrix elements $\langle\tilde{f}|S|\tilde{i}\rangle$ are finite, factorizable in the infra-red factors and directly comparable with the observable inclusive cross sections. More explicitly, one gets

$$\sigma_{\rm incl} \propto |\langle\tilde{f}|S|\tilde{i}\rangle|^2$$
$$= \exp\left\{\int_\lambda \frac{d^3k}{2k} j_\mu(k) j_\mu^*(k)\right\} |\langle f|S|i\rangle_\lambda|^2, \tag{11}$$

where

$$j_\mu(k) = j_\mu^{(i)} + j_\mu^{(f)} = \sum_{\{i,f\}} \alpha \frac{ie}{(2\pi)^{3/2}} \frac{\eta_\alpha p_\mu^{(\alpha)}}{(p^{(\alpha)}k)}, \tag{12}$$

for $k \leqslant \Delta\omega$, and zero otherwise. In eq. (11) the parameter λ has been introduced consistently as the lower limit of the permitted energies of real and virtual photons and has to be finally set to zero. From eq. (11), one finds [‡1]

$$\sigma_{\rm incl} = \sigma_0 \exp\left\{2\frac{\alpha}{\pi}\left[\left(\ln\frac{-q^2}{m^2} - 1\right)\ln\frac{\Delta\omega}{E} + \frac{3}{4}\ln\frac{-q^2}{m^2}\right]\right\}. \tag{13}$$

[‡1] So far the conservation of total energy ($\Sigma k_n \leqslant \Delta\omega$) has not been taken into account. It can be shown (see the second reference of ref. [4]) that this only introduces a correcting factor of order α^2.

Let us observe that the factor $\frac{3}{4}\ln(-q^2/m^2)$ in the right-hand side of eq. (13) has been extracted from the asymptotic elastic form factor given in ref. [11] in the form

$$F^\infty\left(\ln\frac{-q^2}{m^2}, \alpha\right) \sim \exp\left\{-\frac{\alpha}{4\pi}\left[\ln^2\left(\frac{-q^2}{m^2}\right) - 3\ln\left(\frac{-q^2}{m^2}\right)\right] + O(\alpha)\right\}.$$

Let us introduce now in the formalism the collective effect of hard and collinear photons which form, in the limit $m \to 0$, the set of states degenerate to the initial and final particles. Define the new states

$$|\tilde{\tilde{i}}, f\rangle = e^{i\Lambda_{i,f}^{\rm hard}}|\tilde{i}, f\rangle, \tag{14}$$

where $\Lambda_{i,f}^{\rm hard}$ is obtained from eq. (10) with the substitution $j_\mu^{(i,f)} \to j_\mu^{{\rm hard}\,(i,f)}$. The new currents $j_\mu^{\rm hard}(k)$ differ from the old ones $j_\mu(k)$, given by eq. (12), for the different domain of integration, i.e., $\Delta\omega < k < E$ and $\theta_k \leqslant \delta$, where θ_k is the angle between $\boldsymbol{k}$ and $\boldsymbol{p}_a$. In addition a factor $\{1 + [1 - (k/E)]^2\}/2$ is introduced to correct the simple $1/k$ spectrum which appears in any product $j_{\alpha\mu}^{(k)} j_{\beta\mu}^{(k)}$, to account for the hard character of the photons [‡2]. Then we obtain

$$|\langle\tilde{\tilde{f}}|S|\tilde{\tilde{i}}\rangle|^2$$
$$= \exp\left\{\int \frac{d^3k}{2k}[j_\mu^{\rm hard}(k) j_\mu^{\rm hard}(k)^*]\right\} |\langle\tilde{f}|S|\tilde{i}\rangle|^2$$
$$= \exp\left\{\int \frac{d^3k}{2k}[j_\mu^{\rm hard}(k) j_\mu^{\rm hard}(k)^*] + \int_\lambda \frac{d^3k}{2k}[j_\mu(k) j_\mu^*(k)]\right\} |\langle f|S|i\rangle_\lambda|^2. \tag{15}$$

For electron scattering in an external potential we have

$$\int \frac{d^3k}{2k} j_\mu^{\rm hard}(k) j_\mu^{\rm hard}(k)^* = 2\frac{\alpha}{\pi}\left[\left(\ln\frac{E^2\delta^2}{m^2} - 1\right)\left(\ln\frac{E}{\Delta\omega} - \frac{3}{4}\right) + O\left(\frac{\Delta\omega}{E}\right)\right]. \tag{16}$$

[‡2] A more formal derivation of this correction factor for hard photon and gluon emission can be easily found in the literature. See, for example, ref. [12].

Then we obtain for $\sigma_{\text{super}} \propto |\langle \tilde{\tilde{f}}|S|\tilde{\tilde{i}}\rangle|^2$ the following result:

$$\sigma_{\text{super}} = \sigma_0 \exp\left\{2\frac{\alpha}{\pi}\left[\left(\ln\frac{-q^2}{m^2} - 1\right)\ln\frac{\Delta\omega}{E} + \tfrac{3}{4}\ln\frac{-q^2}{m^2}\right]\right\}$$

$$\times \exp\left\{2\frac{\alpha}{\pi}\left(\ln\frac{E^2\delta^2}{m^2} - 1\right)\left(\ln\frac{E}{\Delta\omega} - \frac{3}{4}\right)\right\}, \tag{17}$$

or

$$\sigma_{\text{super}} = \sigma_0 \exp\left\{2\frac{\alpha}{\pi}\left[\ln\frac{-q^2}{k^2_{\perp\max}} \cdot \ln\frac{\Delta\omega}{E} + \tfrac{3}{4}\ln\frac{-q^2}{k^2_{\perp\max}}\right]\right\}. \tag{18}$$

This result explicitly shows that the matrix elements of the new states are free of mass singularities, as stated in the beginning. Furthermore, comparing eqs. (13) and (18) it follows that the final result can be simply obtained from σ_{incl} by the substitution $(1/\gamma) \equiv (m/E) \to \delta$.

So far our results have been expressed in terms of α, the physical electron charge which enters in the definition of the classical currents associated to the coherent state. However, the usual measuring procedure of the charge fails in the limit $m \to 0$, and one has to perform an off-shell renormalization. Furthermore, the emission and reabsorption of electron–positron pairs also become singular in the limit $m \to 0$, and have to be considered explicitly. Therefore one is naturally led to introduce an effective coupling constant in the definition of the classical currents which appears in the coherent states. This simple modification in the formalism automatically reproduces the results previously obtained from the renormalization group equations, as shown below.

The hard component is then given by

$$\int \frac{d^3k}{2k}\, j^{\text{hard}}_\mu(k)\, j^{\text{hard}}_\mu(k)^*$$

$$\approx \int_{m^2}^{k^2_{\perp\max}} \frac{dk^2_\perp}{k^2_\perp} \int_{\Delta\omega/E}^{1} \frac{dx}{x}\,[1 + (1-x)^2]\,\frac{\alpha_{\text{eff}}(M^2, k_\perp)}{\pi} \tag{19}$$

$$\approx \frac{4}{\beta_1}\left(\ln\frac{\Delta\omega}{E} + \frac{3}{4}\right)\ln\left[1 - \frac{\alpha(M^2)}{2\pi}\beta_1 \ln\left(\frac{k^2_{\perp\max}}{m^2}\right)\right],$$

where the effective coupling constant has been defined as

$$\alpha_{\text{eff}}(M^2, k_\perp) = \frac{\alpha(M^2)}{1 - (\alpha(M^2)/2\pi)\beta_1 \ln(k^2_\perp/M^2)}, \tag{20}$$

and $\alpha(M^2)$ is the off-shell coupling constant. A similar treatment of the soft corrections, with $-q^2 \leqslant k^2_\perp < m^2$ then easily leads to eq. (8) instead of eq. (18).

The above analysis shows that the coherent state method gives the same results of the renormalization group equations, providing in addition a very simple physical description of the massless problem.

The generalization of the above results to QCD is straightforward, given the formalism developed in ref. [7] for the soft problem. In fact, for a quark in a singlet potential, the inclusive cross section was found,

$$\sigma_{\text{incl}} \propto |\langle \tilde{f}|S|\tilde{i}\rangle|^2$$

$$= \exp\left\{\int \frac{d^3k}{2k}\,[j^c_\mu(k)\, j^c_\mu(k)^*]\, \bar{g}^2_{\text{YM}}(k)\right\} \langle f|S|i\rangle\,, \tag{21}$$

where $j^c_\mu(k)$ is the coloured generalization of eq. (12), $\bar{g}_{\text{YM}}(k)$ is the effective coupling constant for pure Yang–Mills theory and the right-hand side of eq. (21) is free from infra-red singularities. At the leading logarithm approximation the finite corrections in eq. (21) coming after the cancellation of infra-red singularities can be written as

$$\exp\left\{-\frac{c_F}{2\pi^2}\int_{\Delta\omega}^{E}\frac{dk}{k}\int_{m^2}^{-q^2}\frac{dk^2_\perp}{k^2_\perp}\,\bar{g}^2_{\text{YM}}(k_\perp)\right\}, \tag{22}$$

where we have explicitly introduced a dependence on $k_\perp$ of the effective coupling constant. This dependence is not evident in the results of ref. [7], when only a leading ln k behaviour of $\bar{g}_{\text{YM}}(k)$ was found from the comparison with perturbative calculations. On the other hand a $k_\perp$ dependence of $\bar{g}$ is strongly suggested from the previous results in QED (eq. (20)) and forced by the requirement of cancelling the m^2 singularities between the soft corrections (22) and the hard contributions, analogous to eq. (19), which have been shown [13] to depend explicitly on $\bar{g}(k_\perp)$.

Furthermore, in the limit $m \to 0$, quark loops have to be taken also into account, in addition to pure gluon loops and the effective coupling constant for a pure Yang–Mills theory which appears in eq. (22) has to be replaced by the full $\bar{g}(k_\perp)$.

The hard gluon corrections are obtained from the

Volume 79B, number 4,5 PHYSICS LETTERS 4 December 1978

obvious generalization of eqs. (15) and (19) where β_1 is replaced by the complete β function (quarks + gluons). Then the m^2 singularities exactly cancel between soft and hard contributions and the superinclusive cross section for the scattering of a quark-jet ($\equiv |q\rangle + |q+g\rangle + \ldots + |q + q\bar{q}\rangle + \ldots$) in an external potential becomes, as in eq. (8),

$$\sigma_{\mathrm{super}} = \sigma_0 \exp \left\{ -2c_{\mathrm{F}} \times \int_{\Delta\omega/E}^{1} \frac{\mathrm{d}x}{x} \left[\frac{1+(1-x)^2}{2} \right] \int_{E^2\delta^2}^{-q^2} \frac{\mathrm{d}k_\perp^2}{k_\perp^2} \frac{\bar{\alpha}(k_\perp)}{\pi} \right\} \tag{23}$$

$$= \sigma_0 \exp \left\{ 4 \frac{c_{\mathrm{F}}}{\beta} \left(\ln \frac{\Delta\omega}{E} + \frac{3}{4} \right) \ln \left[1 + \frac{\bar{\alpha}(q^2)}{2\pi} \beta \ln \frac{-q^2}{E^2\delta^2} \right] \right\},$$

where $\beta = -\frac{11}{6} N_{\mathrm{c}} + \frac{2}{3} N_{\mathrm{f}}/2$ and $c_{\mathrm{F}} = (N_{\mathrm{c}}^2 - 1)/2N_{\mathrm{c}}$. In the case of e^+e^- annihilation into two jets, the result of Sterman and Weinberg [14] is simply given by the first-order expansion of eq. (23) ($-q^2 \to 4E^2$). The generalization to other hard processes will be discussed elsewhere.

To summarize, we have studied the problem of mass singularities within the formalism of coherent states. The introduction of the new states, which account both for soft and hard radiation effects, provides one with matrix elements which are free from infra-red and mass singularities. In QED our results coincide with those obtained using the renormalization group equations. The generalization to QCD agrees with low-order calculations and provides a very simple tool to evaluate the finite corrections to hard processes.

We are grateful to R. Meuldermans, Y. Srivastava and G. Veneziano for useful discussions. We wish also to thank G. Veneziano for a critical reading of the manuscript. One of us (M.G.) gratefully acknowledges the warm hospitality enjoyed at the CERN Theoretical Physics Division, where part of this work was done.

References

[1] See, e.g.: D.R. Yennie, S.C. Frautschi and H. Suura, Ann. Phys. 13 (1961) 379.
[2] F. Bloch and A. Nordsiek, Phys. Rev. 52 (1937) 54.
[3] T. Kinoshita, J. Math. Phys. 3 (1962) 650;
T.D. Lee and M. Nauenberg, Phys. Rev. 133 (1964) 1549.
[4] V. Chung, Phys. Rev. B140 (1965) 1110;
M. Greco and G. Rossi, Nuovo Cimento 50 (1967) 168;
T. Kibble, J. Math. Phys. 9 (1968) 315; Phys. Rev. 173 (1968) 1527; 174 (1968) 1882; 175 (1968) 1624;
for more recent work, see the review by: N. Papanicolau, Phys. Rep. 24C (1976) 229.
[5] For a complete list of references, see: W. Marciano and H. Pagels, Phys. Rep. 36C (1978) 137.
[6] J.M. Cornwall and G. Tiktopoulos, Phys. Rev. D13 (1976) 3370; D15 (1977) 2937;
T. Kinoshita and A. Ukawa, Phys. Rev. D15 (1977) 1956; D16 (1977) 332.
[7] M. Greco, F. Palumbo, G. Pancheri-Srivastava and Y. Srivastava, Phys. Lett. 77B (1978) 282.
[8] H.D. Politzer, Nucl. Phys. B129 (1977) 301;
C.T. Sachrajda, CERN Preprint TH. 2416 (1977);
D. Amati, R. Petronzio and G. Veneziano, CERN Preprint TH. 2470 (1978).
[9] E.C.G. Stueckelberg and A. Peterman, Helv. Phys. Acta 26 (1953) 499;
M. Gell-Mann and F. Low, Phys. Rev. 95 (1954) 1300;
N. Bogoliubov and D.V. Shirkov, Nuovo Cimento III (1956) 5.
[10] K.E. Eriksson, in: Cargèse Lectures in Physics, ed. M. Levy (Gordon and Breach, 1967).
[11] C.P. Korthals Altes and E. de Rafael, Nucl. Phys. B106 (1976) 237.
[12] G. Altarelli and G. Parisi, Nucl. Phys. B126 (1977) 298.
[13] Yu.L. Dokshitzer, Leningrad Preprint No. 330 (1977).
[14] G. Sterman and S. Weinberg, Phys. Rev. Lett. 39 (1977) 1436.

Volume 92B, number 1,2 PHYSICS LETTERS 5 May 1980

LARGE INFRA-RED CORRECTIONS IN QCD PROCESSES

G. CURCI and M. GRECO [1]
CERN, Geneva, Switzerland

Received 29 November 1979

The effect of soft gluon emission in hard processes is examined to all orders in QCD. Large corrections have been found, which sensibly modify the parton model results, and show the relevance of the appropriate kinematical limits in leptoproduction and in the Drell–Yan process.

In this letter we study the problem of soft corrections in hard processes, by using the approach developed in ref. [1], based on the combined use of renormalization group techniques and the formalism of coherent states, which has been applied to discuss jet physics in QCD [2].

In particular, we address ourselves to the question raised in ref. [3], where it has been observed that the soft behaviour of the theory plays an important role in the evaluation of the corrections to the leading order results.

Our results confirm to all orders in perturbation theory, in the double logarithm approximation including the effects of the running coupling constant, the importance of exact kinematical constraints in the various processes as recently emphasized by Parisi [4]. We have explicitly considered deep inelastic electroproduction and the Drell–Yan process [5].

Let us first discuss leptoproduction in the $x \simeq 1$ region. It is known that in that limit the parton densities are also exponentiated, in addition to the usual exponentiation of the moments. In fact one finds [6–8], for the valence densities, upon which we concentrate from now on,

$$q(x, \xi) \simeq \frac{\exp(\frac{3}{4} - \gamma_E)C_F\xi}{\Gamma(C_F\xi)} (1-x)^{C_F\xi - 1}, \tag{1}$$

where

$$\xi(k_{\perp\,\max}^2) = \int_{\lambda^2}^{k_{\perp\,\max}^2} \frac{dk_\perp^2}{k_\perp^2} \frac{\alpha(k_\perp)}{\pi}, \tag{2}$$

and $k_{\perp\max}^2 \simeq Q^2$. This result is valid for $[\alpha(Q^2)/\pi]\ln[1/(1-x)] \ll 1$ and $[\alpha(Q^2)/\pi]\ln Q^2 \lesssim 1$. Eq. (1) coincides, as can be easily seen, with the energy distribution of the radiation obtained in the coherent state formalism, taking into account the constraint of energy conservation [9]

$$\frac{dP}{d\omega} = \frac{1}{2\pi} \int_{-\infty}^{\infty} db\, e^{ib\omega} \exp\left\{ C_F \int_0^1 dz \frac{1+(1-z)^2}{2z} \xi(Q^2) [e^{-ibz} - 1] \right\}, \tag{3}$$

where $\omega = 1 - x$ is the fraction of energy carried out by the soft gluon radiation.

More generally, in order to show explicitly the equivalence of the nth moment of the distribution (3) with the

[1] Permanent address: Laboratori Nazionali INFN, Frascati, Italy.

Volume 92B, number 1,2 PHYSICS LETTERS 5 May 1980

conventional parton density moments for large n, let us rewrite eq. (3) in the form

$$\frac{\mathrm{d}P}{\mathrm{d}x} = q(x, Q^2) = \frac{1}{2\pi\mathrm{i}} \int_{-\mathrm{i}\infty}^{\mathrm{i}\infty} \mathrm{d}n\, \mathrm{e}^{n(1-x)} \exp\left\{ \frac{C_F}{2} \int_0^1 \mathrm{d}y\, \frac{1+y^2}{1-y}\, \xi(Q^2) [\mathrm{e}^{-n(1-y)} - 1] \right\}$$

$$\simeq \frac{1}{2\pi\mathrm{i}} \int_{-\mathrm{i}\infty}^{\mathrm{i}\infty} \mathrm{d}n\, x^{-n} \exp\left[\frac{C_F}{2} \int_0^1 \mathrm{d}y\, \frac{1+y^2}{1-y} \xi(Q^2)(y^n - 1) \right] . \tag{4}$$

This result explicitly shows that the moments of the distribution (3) coincide with the conventional moments of parton densities in the large-n limit.

Let us discuss now the very soft region, when $[\alpha(Q^2)/\pi] \ln [1/(1-x)] \lesssim 1$. In order to study the soft singularities, we use as the infra-red regulator the gluon mass $=\lambda$ with zero quark mass. This regularization scheme makes clearer the relevance of the kinematical constraints in the transverse momentum phase space in all processes. Similar results can be obtained using the dimensional regularization scheme [‡1]. This choice of regularization will also be convenient below in discussing the large-Q^2 behaviour of the quark form factor.

To first order, the electromagnetic quark form factor $F(Q^2)$ is given by $(q^2 = -Q^2 < 0)$

$$F(Q^2) \simeq 1 - (\alpha/4\pi) C_F \ln^2(Q^2/\lambda^2) , \tag{5}$$

in the leading double logarithm approximation. Correspondingly, the real contribution to the valence parton density is

$$q^{\mathrm{real}}(x, Q^2) \simeq \delta(1-x) + \frac{\alpha}{\pi} C_F \frac{1}{1-x} \ln\left[\frac{Q^2}{\lambda^2}(1-x) \right]$$

$$= \delta(1-x) + \frac{\alpha}{\pi} C_F \left\{ \frac{1}{1-x} \ln\left[\frac{Q^2}{\lambda^2}(1-x) \right] \right\}_+ + \frac{\alpha}{\pi} C_F \delta(1-x) \int_0^{1-\lambda^2/Q^2} \frac{\mathrm{d}y}{1-y} \ln\left[\frac{Q^2}{\lambda^2}(1-y) \right] . \tag{6}$$

From eqs. (5) and (6) it follows that the $\ln^2\lambda$ singularity is cancelled by adding the real and virtual contribution to $q(x, Q^2)$, and in addition the first moment is equal to one, as requested by the Adler sum rule [10]. In order to show the exponentiation of the $1/(1-x) \ln (1-x)$ singularity, we will require below the same cancellation of the $\ln^2\lambda$ singularity between real and virtual contributions, after use of the appropriate expression of the quark form factor, demanding in addition the fulfilment of the Adler sum rule.

By ignoring the effect of the running coupling constant, which will be discussed later, and using the result for the exponentiation of the leading $\ln^2(Q^2/\lambda^2)$ singularity of the form factor [11], we obtain for the nth moment

$$q^{(m)}(Q^2) \simeq \exp\left\{ \frac{\alpha}{\pi} C_F \int_0^{1-\lambda^2/Q^2} \frac{\mathrm{d}y}{1-y} \ln\left[\frac{Q^2}{\lambda^2}(1-y) \right] y^{n-1} \right\} \exp\left(-\frac{\alpha}{2\pi} C_F \ln^2 \frac{Q^2}{\lambda^2} \right)$$

$$= \exp\left\{ \frac{\alpha}{\pi} C_F \int_0^1 \frac{\mathrm{d}y}{1-y} \ln\left[\frac{Q^2}{\lambda^2}(1-y) \right] (y^{n-1} - 1) \right\} . \tag{7}$$

The same result follows directly in the coherent state approach, by simple kinematical bounds in the $k_\perp$ phase space in eq. (2), namely:

$$\frac{\mathrm{d}P}{\mathrm{d}x} = q(x, Q^2) \simeq \frac{1}{2\pi} \int_{-\infty}^{\infty} \mathrm{d}b\, \mathrm{e}^{\mathrm{i}b(1-x)} \exp\left\{ C_F \int_0^1 \frac{\mathrm{d}z}{1-z} \int_{\lambda^2}^{k_{\perp\max}^2 = Q^2(1-z)} \frac{\mathrm{d}k_\perp^2}{k_\perp^2} \frac{\alpha(k_\perp)}{\pi} [\mathrm{e}^{-\mathrm{i}b(1-z)} - 1] \right\} , \tag{8}$$

‡1 See, for example, ref. [3].

which shows, by the argument given above, that the nth moment of eq. (8) coincides with eq. (7) for large n and $\alpha(k_\perp) = \alpha$ = constant.

We now take into account the effect of the running coupling constant. In our regularization scheme and with on-shell normalization conditions, an explicit two-loop calculation [12] shows that the quark form factor satisfies the Callan–Symanzik equation:

$$\left[\lambda\frac{\partial}{\partial\lambda} + \beta\frac{\alpha}{\pi}\alpha\frac{\partial}{\partial\alpha} - \frac{\alpha}{\pi}C_F\left(\ln\frac{Q^2}{\lambda^2} - 3\right)\right] F(\alpha, Q^2/\lambda^2) = 0\,, \tag{9}$$

where $\beta = -\frac{1}{2}(\frac{11}{3}N_c - \frac{2}{3}N_f)$. Assuming the validity of eq. (9) to all orders, we obtain in the leading logarithm approximation the solution

$$F(\alpha, Q^2/\lambda^2) = \exp\left[-\frac{C_F}{2\beta}\ln\left(1 - \frac{\alpha}{2\pi}\beta\ln\frac{Q^2}{\lambda^2}\right)\ln\frac{Q^2}{\lambda^2}\right]. \tag{10}$$

A similar result has been found using a different regularization scheme in ref. [13].

In analogy with eq. (7), we then find

$$q^{(n)} \simeq \exp\left[C_F\int_0^{1-\lambda^2/Q^2}\frac{dz}{1-z}\int_{\lambda^2}^{Q^2(1-z)}\frac{dk_\perp^2}{k_\perp^2}\frac{\alpha(k_\perp^2)}{\pi}z^{n-1}\right]|F|^2$$

$$= \exp\left[C_F\int_0^1\frac{dz}{1-z}(z^{n-1}-1)\int_{\lambda^2}^{Q^2(1-z)}\frac{dk_\perp^2}{k_\perp^2}\frac{\alpha(k_\perp^2)}{\pi}\right]. \tag{11}$$

An intuitive derivation of this result simply follows in the coherent state picture, by taking into account the $k_\perp$ dependence of α in eq. (8). So our expression (8) replaces eq. (1) in the very near kinematical limit of $x \simeq 1$. This result has been also obtained by Gribov and Lipatov [6], and Dokshitzer [7] by a careful diagrammatic analysis. In particular, in ref. [7] an explicit expression for $q(x, Q^2)$ is given by evaluating eq. (8) by saddle point techniques.

We note also that the Altarelli–Parisi [14] evolution equation for the nonsinglet quark density becomes in this limit

$$\frac{dq(x, Q^2/Q_0^2)}{d\ln(Q^2/Q_0^2)} = \frac{1}{2\pi}\int_x^1\frac{dy}{y}\,[P(x/y)\bar{\alpha}((Q^2/Q_0^2)(1-x/y))]_+\,q(y, Q^2/Q_0^2)\,, \tag{12}$$

where Q_0^2 is a normalization scale, $P(z) = C_F[(1+z^2)/(1-z)]$ and we have implicitly absorbed the mass singularities in the distribution of partons inside the hadron.

We now discuss the implications of our results in the Drell–Yan process. As is well known, in the naive parton model one has

$$\frac{d\sigma^{DY}}{dQ^2} = \frac{4\pi\alpha^2}{9SQ^2}\int\frac{dx_1}{x_1}\frac{dx_2}{x_2}\left[\sum_i e_i^2 q_i(x_1)\bar{q}_i(x_2) + (1\leftrightarrow 2)\right]\delta(1-z)\,, \tag{13}$$

where $\sqrt{S}$ is the invariant mass of the hadronic system, $z = \tau/x_1x_2$ and $\tau = Q^2/S$.

By taking into account the soft gluon emission at all orders and omitting for simplicity the contribution of gluons in the initial state, the delta function in eq. (13) is replaced in the coherent state approach, in the limit $z \simeq 1$, by

$$f(z, Q^2) \simeq \left\{\frac{1}{2\pi}\int db\, e^{ib(1-z)}\exp\left[2C_F\int_0^1 dy\,\frac{1}{1-y}\int_{\lambda^2}^{Q^2(1-y)^2}\frac{dk_\perp^2}{k_\perp^2}\frac{\alpha(k_\perp)}{\pi}\,[e^{-ib(1-y)}-1]\,\exp\{[\alpha(Q^2)/2\pi]C_F\pi^2\}\,, \tag{14}$$

in analogy to eq. (8).

Let us briefly comment on our result. The factor $2C_F$ is simply related to the emission of gluons from two legs

Volume 92B, number 1,2 PHYSICS LETTERS 5 May 1980

in the initial state and the upper limit $Q^2(1-y)^2$ in the integral is due to the kinematical bound $k_{\perp\max}^2 = Q^2(1-y)^2/y \simeq Q^2(1-y)^2$. In addition, the factor $\exp\{[\alpha(Q^2)/2\pi]C_F\pi^2\}$ is due to the continuation of the Q^2 dependence in the form factor from space-like to time-like regions [4] [see eq. (5) in exponentiated form].

Eqs. (13) and (14) can now be rewritten as

$$\frac{d\sigma^{DY}}{dQ^2}=\frac{4\pi\alpha^2}{9SQ^2}\int\frac{dx_1}{x_1}\frac{dx_2}{x_2}\left[\sum_i e_i^2 q_i(x_1,Q^2)\bar{q}_i(x_2,Q^2)+(1\to 2)\right]\tilde{f}(z,Q^2)\,, \tag{15}$$

with

$$\tilde{f}(z,Q^2)=\left\{\frac{1}{2\pi}\int db\, e^{ib(1-z)}\exp\left[2C_F\int_0^1\frac{dy}{1-y}\int_{Q^2(1-y)}^{Q^2(1-y)^2}\frac{dk_\perp^2}{k_\perp^2}\frac{\alpha(k_\perp)}{\pi}\left(e^{-ib(1-y)}-1\right)\right]\right\}\exp\{[\alpha(Q^2)/2\pi]C_F\pi^2\}\,, \tag{16}$$

having explicitly introduced the Q^2 dependence in the parton densities in deep inelastic scattering. The equivalence of eqs. (15) and (16) with eqs. (13) and (14) follows directly by taking the τ^n moments of $d\sigma/dQ^2$ in both equations and using eq. (8) for the Q^2 dependence of the parton densities. It is easy to check that our final formulae (15), (16) agree with the perturbative calculation of ref. [3] in the limit $z \to 1$.

We note that the term $\exp\{[\alpha(Q^2)/2\pi]C_F\pi^2\}$ in eq. (16) plays an important role in renormalizing the naive Drell–Yan cross section by about a factor of two at present energies and should therefore be taken into account in comparison with experimental data. More detailed phenomenological implications of our results will be discussed elsewhere.

We now comment briefly on the transverse momentum distributions of the μ pairs in the Drell–Yan process. In our framework the $k_\perp$ distributions can be simply derived in analogy with eq. (8) taking into account the constraint of transverse momentum conservation on the multiple gluon emission, as explicitly found in ref. [15]. They coincide with those obtained in ref. [2], once applied to this particular process.

Our results can be summarized as follows. The formalism of coherent states has been shown to be quite appropriate for studying the effects of soft gluon bremsstrahlung at all orders. Large corrections have been found which explicitly show the importance of the soft regions in various processes. An intuitive explanation of these corrections follows from simple kinematical considerations. Similar arguments apply also to the e^+e^- annihilation.

We acknowledge useful discussions with R. Petronzio and G. Veneziano.

References

[1] G. Curci and M. Greco, Phys. Lett. 79B (1978) 406.
[2] G. Curci, M. Greco and Y. Srivastava, Phys. Rev. Lett. 43 (1979) 834; Nucl. Phys. B159 (1979) 451.
[3] G. Altarelli, R.K. Ellis and G. Martinelli, Nucl. Phys. B157 (1979) 461.
[4] G. Parisi, Frascati preprint LNF-79/71 (P) (1979).
[5] S.D. Drell and T.M. Yan, Phys. Rev. Lett. 25 (1970) 316.
[6] V.N. Gribov and L.N. Lipatov, Sov. J. Nucl. Phys. 15 (1972) 458, 675.
[7] Yu.L. Dokshitser, Sov. J. Nucl. Phys. 46 (1977) 641.
[8] K. Konishi, A. Ukawa and G. Veneziano, Nucl. Phys. B157 (1979) 45.
[9] E. Etim, G. Pancheri and B. Touschek, Nuovo Cimento LIB (1967) 276;
G. Curci, M. Greco, Y. Srivastava and B. Stella, Phys. Lett. 88B (1979) 147;
M. Ramon-Medrano, G. Pancheri-Srivastava and Y. Srivastava, Frascati preprint LNF-79/41(P) (1979).
[10] S.L. Adler, Phys. Rev. 143 (1965) 1144.
[11] J.M. Cornwall and G. Tiktopoulos, Phys. Rev. D13 (1976) 3370.
[12] G. Curci, in preparation.
[13] R. Coqueraux and E. de Rafael, Phys. Lett. 74B (1978) 105.
[14] G. Altarelli and G. Parisi, Nucl. Phys. B126 (1977) 298.
[15] G. Parisi and R. Petronzio, Nucl. Phys. B154 (1979) 427.

VOLUME 43, NUMBER 12 PHYSICAL REVIEW LETTERS 17 SEPTEMBER 1979

Coherent Quark-Gluon Jets

G. Curci
CERN, Geneva, Switzerland

and

M. Greco and Y. Srivastava[(a)]
Laboratori Nazionali di Frascati, Istituto Nazionale di Fisica Nucleare, Frascati, Italy

(Received 5 March 1979)

With use of the coherent-state formalism, general expressions are presented for arbitrary massless quark and gluon jet cross sections. Also, normalized probability distributions are given for the jet transverse momentum which take due account of correlations imposed by the momentum conservation. These results are used to discuss successfully the SPEAR data of Hanson *et al*. between 3 and 7.8 GeV.

Much attention has been directed recently to study theoretically the jet structure in quantum chromodynamics (QCD) perturbation theory.[1] In the present work, we describe the results of an approach based on the coherent-state formalism.[2,3] A simple formula is obtained for a general jet cross section in terms of an infinite number of gluons and massless quarks. We also obtain probability distributions for the jet transverse momentum which include correlations induced by the momentum conservation. A phenomenological extrapolation of our jet $K_\perp$ distribution in the nonperturbative region is then presented. This assumes that the hadronic transverse-momentum distribution follows that of the QCD radiation[4] and uses a parametrization of $\bar{\alpha}(k_\perp)$ obtained in an earlier work.[5] We find excellent agreement with the accurate SPEAR data for the two-jet process $e^+e^- \to q\bar{q} \to$ hadrons, for total energies 3–7.8 GeV.[6]

In Ref. 2, coherent states $|\bar{i}\rangle$, consisting of an indefinite number of soft gluons, corresponding to a "pure" state $|i\rangle$ were constructed to obtain ir-finite inclusive cross sections in QCD. For example, for a quark in a color-singlet potential, the inclusive cross section was found to be[2]

$$\sigma_{\rm inc} \propto |\langle \bar{f}|S|\bar{i}\rangle|^2 = \exp\{\textstyle\int d^3k\,(2k)^{-1}[j_\mu{}^a(k)^\dagger]\bar{g}_{\rm YM}{}^2(k)\}\,|\langle f|S|i\rangle|^2, \tag{1}$$

where $j_\mu{}^a(k)$ is the classical color current operator and $\bar{g}_{\rm YM}(k)$ is the effective coupling constant for pure Yang-Mills theory. The finite corrections in Eq. (1) after the cancellation of ir singularities can be written as

$$\exp\left[-\frac{C_F}{2\pi^2}\int_{\Delta\omega}^{E}\left(\frac{dk}{k}\right)\int_{m^2}^{-q^2}\frac{dk_\perp{}^2}{k_\perp{}^2}\bar{g}_{\rm YM}{}^2(k)\right], \tag{2}$$

where $C_F = (N_C{}^2-1)/2N_C$ for SU(N_C) color, m and $\Delta\omega$ are the mass and energy loss of the quark, and q^2 is the momentum transfer.

To discuss the massless limit $m\to 0$ new coherent states have to be defined[3] which take into account the collective effect of hard and collinear gluons, i.e., the set of states degenerate with the initial and final states.[7] As discussed in Ref. 3 the mass singularities exactly cancel between soft and hard contributions provided $g_{\rm YM}(k)$ in the above Eq. (2) is replaced by the full coupling constant $\bar{g}(k_\perp)$, since, as $m\to 0$, quark loops in addition to the gluons must be included.

Thus, for $e^+e^-\to\bar{q}q$ we have ir- and mass-singularity-free "super" inclusive cross sections given by

$$d\sigma_{\rm super}{}^{(2q)} = d\sigma_0 \exp\left[-\frac{1}{\pi^2}\int_{\Delta\omega/E}^{1}dx\,P_{gq}(x)\int_{k_{\perp 1}{}^2}^{k_{\perp 2}{}^2}\frac{(d^2k_\perp)}{k_\perp{}^2}\bar{\alpha}(k_\perp)\right], \tag{3}$$

where the gluon distributions due to the quarks is[8]

$$P_{gq}(x) = C_F[1+(1-x)^2]/x. \tag{4}$$

In Eq. (3) $k_{\perp 1} = E\delta$, $k_{\perp 2} = Q/2 = E$, E is the quark energy, δ is the half angle of the jet cone, and $d\sigma_0$ is the "pointlike" cross section.

The ratio $d\sigma_{\rm super}/d\sigma_0$ in Eq. (3) gives the probability of finding a fraction $\epsilon = \Delta\omega/Q$ of the total energy Q outside a pair of oppositely directed cones of half angle δ. First order expansion of Eq. (5) directly gives the result of Sterman and Weinberg[1] up to terms of order $\epsilon\ln\delta$. A more accurate kinematical analysis shows that the upper limit in x should be restricted to $1-\Delta\omega/E=1-2\epsilon$. This gives

$$I_q(\epsilon)\equiv\int_{2\epsilon}^{1-2\epsilon}dx\,P_{gq}(x)=-C_F[2\ln2\epsilon+\tfrac{3}{2}-2\epsilon+4\epsilon^2+O(\epsilon^3)], \tag{5}$$

in agreement with Stevenson.[1] From now, we exponentiate with the right kinematics and certainly our results will remain valid for small ϵ but they may not account for all ϵ terms.

The generalization of Eq. (3) to gluon jets is obtained as follows. The probability $P_{gq}(x)$ in Eq. (4) has to be replaced by

$$P_{gg}(x)+N_fP_{qg}(x)=N_C\left[\frac{x}{1-x}+\frac{1-x}{x}+x(1-x)\right]+N_f\,\frac{x^2+(1-x)^2}{2}, \tag{6}$$

corresponding to the sum of probabilities that a gluon radiates gluons radiates gluons or $N_f(q\bar{q})$ pairs. Then Eq. (5) is replaced by

$$\begin{aligned}I_g(\epsilon)&\equiv\int_{2\epsilon}^{1-2\epsilon}dx[P_{gg}(x)+N_fP_{qg}(x)]\\&=-N_C\left[2\ln2\epsilon+\left(\frac{11}{6}-\frac{1}{3}\frac{N_f}{N_C}\right)-4\epsilon\left(1-\frac{N_f}{2N_C}\right)+8\epsilon^2\left(1-\frac{N_f}{2N_C}\right)+O(\epsilon^3)\right].\end{aligned} \tag{7}$$

The logarithmic and constant terms (in ϵ) in Eq. (7) agree with the explicit perturbative calculations of Shyzuya and Tye, Einhorn and Weeks, and Smilga and Vysotsky.[1] Finite order terms in ϵ have, as yet, not been calculated in perturbation theory. Notice that our distributions in ϵ for $I_q(\epsilon)$ and $I_g(\epsilon)$ have been obtained without any recourse to singular distributions.[8] This difference arises because we do not integrate to $x=1$.

The above results generalize for n_q and n_g quark and gluon jets, respectively, to

$$d\sigma_{\text{super}}^{(n_q,n_g)}=d\sigma_0\exp\left\{-\frac{1}{\pi}[n_qI_q(\epsilon)+n_gI_g(\epsilon)]\int_{k_{\perp1}}^{k_{\perp2}}\left(\frac{dk_\perp}{k_\perp}\right)\bar{\alpha}(k_\perp)\right\}, \tag{8}$$

where for simplicity the same kinematical limits $k_{\perp1}$ and $k_{\perp2}$ are used for all jets. The curly bracket in Eq. (8) gives the leading terms in ($\ln\delta$). All nonleading terms in δ are included in $d\sigma_0$ and have to be calculated, if need be, for each process separately in perturbation theory. First-order expansion of Eq. (18) agrees with the results of Smilga and Vysotsky.[1]

From now on we only consider $q\bar{q}$ jets in e^+e^- annihilation. From Eq. (3), we can define a probability distribution $dP(K_\perp)$,

$$\int_0^{K_{\perp1}}\left(\frac{dP}{dK_\perp}\right)dK_\perp\equiv\exp\left[-\frac{2}{\pi}I(\epsilon)\int_{K_{\perp1}}^{Q/2}\frac{dk_\perp}{k_\perp}\bar{\alpha}(k_\perp)\right], \tag{9}$$

so that

$$\frac{dP}{dK_\perp}\simeq\frac{2I(\epsilon)}{\pi}\left[\frac{\bar{\alpha}(K_\perp)}{K_\perp}\right]\left[\frac{\bar{\alpha}(Q/2)}{\bar{\alpha}(K_\perp)}\right]^{I(\epsilon)/\pi b}, \tag{10}$$

where $b=(33-2N_f)/12\pi$ for three colors, $I(\epsilon)\equiv Iq(\epsilon)$ and we have used the asymptotic-freedom formula for $\bar{\alpha}(k_\perp)$. Thus, Eq. (10) should be valid only for $\Lambda\ll K_\perp\ll Q/2$.

So far, we have omitted any correlation in transverse momentum, since it was assumed that the individual emissions were statistically independent. We can impose transverse-momentum conservation in a manner similar to that of Greco, Pancheri-Srivastava, and Srivastava.[9] This leads us to the following results, the details of which shall be presented elsewhere. We obtain

$$\frac{d^2P}{d^2K_\perp}=\frac{1}{2\pi}\int_0^\infty x_\perp dx_\perp J_0(x_\perp K_\perp)\exp\left\{-\frac{2I(\epsilon)}{\pi}\int_0^{Q/2}\left(\frac{dk_\perp}{k_\perp}\right)\bar{\alpha}(k_\perp)[1-J_0(x_\perp k_\perp)]\right\}, \tag{11}$$

and

$$P(\epsilon,K_{\perp1})=K_{\perp1}\int_0^\infty dx_\perp J_1(x_\perp K_{\perp1})\exp\left\{-\frac{2}{\pi}I(\epsilon)\int_0^{Q/2}\left(\frac{dk_\perp}{k_\perp}\right)\bar{\alpha}(k_\perp)[1-J_0(x_\perp k_\perp)]\right\}. \tag{12}$$

Equation (12) gives the probability of finding a fraction $1-\epsilon$ of the total energy inside two opposite cones of maximum transverse momentum $K_{\perp 1}$, and replaces our earlier Eq. (3) for $(d\sigma_{\text{super}}/d\sigma_0)$. Similarly, Eq. (11), which is properly normalized to 1 for any fixed ϵ, replaces the uncorrelated emission result [Eq. (10)]. From it various moments can be derived. For example,

$$\langle K_\perp^2\rangle = (I(\epsilon)/\pi)\int_0^{Q^2/4} dk_\perp^2\, \bar{\alpha}(k_\perp), \tag{13}$$

which for large Q becomes

$$\langle K_\perp^2\rangle \simeq \text{const} + \frac{I(\epsilon)}{4\pi} Q^2 \bar{\alpha}(\tfrac{1}{2}Q)\left[1 + O\left(\frac{1}{\ln Q}\right)\right]. \tag{14}$$

The constant term can only be obtained upon a knowledge of $\bar{\alpha}(k_\perp)$ in the nonperturbative region $(K_\perp \lesssim C\Lambda)$. Using the parametrization discussed below, we find it negligible for a wide range of C values ($C \gtrsim 1$). A comparison of quark and gluon jets then gives, for fixed ϵ and Q,

$$\frac{\langle K_\perp^2\rangle_{g\ \text{jet}}}{\langle K_\perp^2\rangle_{q\ \text{jet}}} = \frac{I_g(\epsilon)}{I_q(\epsilon)} \xrightarrow[\epsilon\to 0]{} \frac{2N_C^2}{N_C^2-1}, \tag{15}$$

which predicts gluon jets broader by a factor of two than the quark jets. Similar results have been obtained in the recent literature.[1,4]

A few remarks are in order concerning the validity of the above results, since, as is clear from Eqs. (11-12), the range of the $k_\perp$ integrals extends over a nonperturbative region $(k_\perp \lesssim C\Lambda)$ where the leading logarithmic approximation is questionable. The influence of this region is negligible for $K_\perp \gg C\Lambda$, which is the region of interest for a direct test of QCD, as can be seen by approximating $J_0(z)$ by $\theta(1-z)$ and $J_1(z)$ by $\delta(1-z)$ in Eqs. (11-12) which then reduce to the naive expressions (3) and (10). Numerical results which explicitly verify this fact, as well as giving our predictions for jet physics at the forthcoming very high-energy machines, will be presented elsewhere.

We now present an interesting result obtained by extrapolating smoothly our result (11) to the region $K_\perp \lesssim C\Lambda$. Assuming that the above $K_\perp$ distribution obtained for the radiation of a $q\bar{q}$ jet also describes the $k_\perp$ behavior of a single hadron[4] and using a simple parametrization for $\bar{\alpha}(k_\perp)$, directly suggested[5] by all the data on $R \equiv \sigma(e\bar{e} \to \text{hadrons})/\sigma(e\bar{e}\to\mu\mu)$, we find that Eq. (11) reproduces the experimental data quite well. More in detail, we limit the maximum transverse momentum allowed for a single hadron to a value $\sim Q/2\langle n\rangle$, where $\langle n\rangle$ is the average multiplicity.

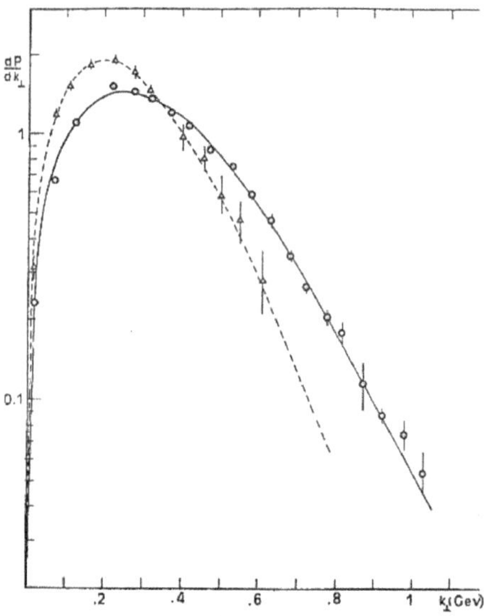

FIG. 1. Normalized $\sigma^{-1}d\sigma/dk_\perp$ vs $k_\perp$ for $Q = 3$ GeV (triangles) and 7.5 GeV (circles) SPEAR single inclusive data from Ref. 6. Our results are given by the dashed line for 3 GeV and by the solid line for 7.5 GeV.

This automatically restricts the $k_\perp$ range in Eq. (11) to values $\lesssim 2$-3 GeV for present and near future e^+e^- beam energies. For $\bar{\alpha}(k_\perp)$ we use the following parametrization:

$$\bar{\alpha}(k_\perp) = \begin{cases} \dfrac{6\pi}{(33-2N_f)\ln C}, & k_\perp \leq C\Lambda \\[2ex] \dfrac{6\pi}{(33-2N_f)\ln(k_\perp/\Lambda)}, & k_\perp \geq C\Lambda, \end{cases} \tag{16}$$

with $C \simeq 4$ and $\Lambda \simeq 0.7$ GeV. This effective $\bar{\alpha}$ has been shown[5] to bring the QCD prediction $R = \sum_s Q_i^2[1+\bar{\alpha}(S)/\pi]$ in excellent agreement with the average value $R_{\text{av}} \equiv \bar{S}^{-1}\int_{S_{\text{th}}} ds\, R_{\text{exp}}(S)$, extracted from *all* data below $\bar{S} \simeq 60$ GeV^2.

With the above modifications in Eq. (11) our predictions are plotted in Fig. (1) for $Q = 3$ and 7.5 GeV, and compared with the SLAC data[6] on $\sigma^{-1}d\sigma/dk_\perp$, suitably normalized to unity. Being unable to extract ϵ directly from the data, we have fixed $I(\epsilon)$ by requiring the same $\langle k_\perp\rangle$ as given by the data at $Q = 7.5$ GeV. This leads to $2\epsilon \simeq 0.025$, which is also used for $Q = 3$ GeV. The data in Fig. (1) agree very well with our expressions and thus support our extrapolation of the

QCD $K_\perp$ distribution to the nonperturbative region.

A useful analytical approximation to Eq. (11), valid for $\bar{\alpha}$ constant, is given by

$$\frac{dP}{dk_\perp} \simeq \frac{2\beta}{a\Gamma(1+\beta/2)}\left(\frac{k_\perp}{a}\right)^{\beta/2} K_{1-\beta/2}\left(\frac{2k_\perp}{a}\right), \qquad (17)$$

where $a = Q/2$, $\beta = 2\bar{\alpha}I(\epsilon)/\pi$, and $K_\nu(z)$ is the modified Bessel function. This result was derived earlier in Ref. 4. Equation (17) gives the same $\langle k_\perp\rangle$ and $\langle k_\perp^2\rangle$ as the original distribution (11).

In conclusion, we have presented a general formalism for quark and gluon jets and their $k_\perp$ distributions. A successful analysis of single-hadron $k_\perp$ distributions at SPEAR ($Q = 3-7.8$ GeV) indicates a rather smooth extrapolation of QCD for predictions to the nonperturbative region as also found[5] for the total cross section.

One of us (Y.S.) would like to thank Dr. Giulia Pancheri-Srivastava for useful discussions. This work was supported, in part, by the U. S. National Science Foundation, Washington, D. C.

[(a)] Permanent address: Northeastern University, Boston, Mass. 02115.

[1] G. Sterman and S. Weinberg, Phys. Rev. Lett. 39, 1436 (1977); H. Politzer, Nucl. Phys. B129, 301 (1977); H. Georgi and M. Machacek, Phys. Rev. Lett. 39, 1237 (1977); E. Farhi, Phys. Rev. Lett. 39, 1587 (1977); P. Stevenson, Phys. Lett. 78B, 451 (1978); K. Shizuya and S.-H. H. Tye, Phys. Rev. Lett. 41, 787 1195(E) (1978); M. Einhorn and B. Weeks, SLAC Report No. SLAC-PUB 2164, 1968 (unpublished); A. Smilga and M. Vysotsky, Institute for Theoretical and Experimental Physics (Moscow) Report No. ITEP-93, 1978 (unpublished); K. Koller and T. Walsh, Nucl. Phys. B140, 449 (1978); G. Fox and S. Wolfram, Phys. Rev. Lett. 41, 1581 (1978); K. Konishi, A. Ukawa, and G. Veneziano, Phys. Lett. 80B, 259 (1979); R. K. Ellis and R. Petronzio, Phys. Lett. 80B, 249 (1979); A. Ali, J. Körner, Z. Kunszt, J. Willrodt, G. Kramer, G. Schieroholz, and E. Pietarinen, to be published.

[2] M. Greco, F. Palumbo, G. Pancheri-Srivastava, and Y. Srivastava, Phys. Lett. 77B, 282 (1978).

[3] G. Curci and M. Greco, Phys. Lett. 79B, 406 (1978).

[4] G. Pancheri-Srivastava and Y. Srivastava, Laboratori Nazionali di Frascati Report No. LNF-78/46(P), 1978 (to be published); G. Pancheri-Srivastava and Y. Srivastava, Northeastern University (Boston) Report No. NUB 2379, 1978 (to be published).

[5] M. Greco, G. Penso, and Y. Srivastava, to be published.

[6] G. Hanson, in *Proceedings of the Thirteenth Recontre de Moriond, Les Arcs-Savoie, France, 1978*, edited by J. Tran Thanh Van (Editions Frontières, France, 1979).

[7] T. Kinoshita, J. Math. Phys. 3, 650 (1962); T. D. Lee and M. Nauenberg, Phys. Rev. B 133, 1549 (1964).

[8] G. Altarelli and G. Parisi, Nucl. Phys. B126, 298 (1977).

[9] M. Greco, G. Pancheri-Srivastava and Y. Srivastava, Nucl. Phys. B101, 234 (1975).

Particle-Gamma Coincidence Study of Gross-Structure Peaks in $^{12}C + ^{12}C$ Inelastic Scattering

L. E. Cannell, R. W. Zurmühle, and D. P. Balamuth
Physics Department, University of Pennsylvania, Philadelphia, Pennsylvania 19104
(Received 1 May 1979; revised manuscript received 16 July 1979)

Particle-γ angular correlations with forty ^{12}C angles for a fixed γ angle have been measured for the reaction $^{12}C(^{12}C, ^{12}C)^{12}C^*$(4.44 MeV) over the energy range $14.5 \leq E_{c.m.} \leq 32$ MeV. Satisfactory agreement with a large body of experimental data is obtained using a simple distorted-wave Born-approximation calculation without introducing the idea of "quasimolecular" resonances.

The recent discovery[1,2] of pronounced periodic structure in the excitation curves for inelastic scattering of two ^{12}C nuclei has stimulated considerable interest. Various interpretations of this periodic structure have been advanced.[2-6] Among these, two extreme points of view may be noted. First, one can assume that these anomalies are manifestations of quasibound 24-nucleon "molecular" states with definite spin and parity. Alternatively, it is possible that the gross structure in the inelastic scattering excitation curves is a phenomenon only indirectly related to the ^{24}Mg system and is primarily the consequence of simple kinematic relationships involving the dominance of surface partial waves.

In several of the theoretical models involving nuclear quasimolecules these states form one or more rotational bands with $E(J)$ proportional to $J(J+1)$; these bands are characterized by a moment of inertia roughly given by two carbon nuclei rotating about each other. The simplicity of such a picture makes it attractive, but if it is to

Nuclear Physics B159 (1979) 451–468

QCD JETS FROM COHERENT STATES

G. CURCI
CERN, Geneva, Switzerland

M. GRECO and Y. SRIVASTAVA *‡
INFN, Laboratori Nazionali di Frascati, Frascati, Italy

Received 27 February 1979

A recently proposed approach to the problem of infrared and mass singularities in QCD, based on the formalism of coherent states, is extended to discuss massless quark and gluon jets. Our results include all leading ($\ln \delta$) terms as well as finite terms in the energy loss ϵ, in addition to the usual $\ln \epsilon$ associated with $\ln \delta$. Our formulae agree with explicit perturbative calculations, whenever available. Explicit expressions for the total K_T distributions are given which take into account transverse-momentum conservation. Predictions are also made for the Q^2 dependence of the mean K_T^2 for quark and gluon jets. Our jet k_T distributions are extrapolated for low k_T and shown to describe with good accuracy the data for $e\bar{e} \to q\bar{q} \to$ hadrons. Numerical predictions are also presented for the forthcoming PETRA, PEP and LEP machines.

1. Introduction

Experimental evidence [1] for jets in e^+e^- annihilation and the observed [2] features of scaling violations in deep inelastic scattering have given great support to the quark parton model and its underlying theory, quantum chromodynamics (QCD). Considerable efforts have been recently devoted to working out quantitative predictions of QCD for hard processes [3]. After the work of Sterman and Weinberg [4], jet phenomena have been considered in great detail. The basic idea is to compute suitable defined jet cross sections which are free of infrared and mass singularities [5]. Much progress has been made in calculating higher-order terms in the leading logarithmic approximation, for various processes. These leading terms have been shown to sum up to an exponential form, for which a very convenient infrared, mass-singularity free coherent-state formalism has been developed [6,7].

Let us briefly summarize the basic idea of this approach. The starting point is

* Permanent address: Northeastern University, Boston, Mass., USA.

‡ Work supported in part by the National Science Foundation, Washington, USA.

the well-known fact that all infrared singularities arising from real and virtual soft quanta have to cancel in the physical cross sections [8,9]. This cancellation occurs to all orders of perturbation theory. The formalism of coherent states, developed in ref. [6], indeed provides one with matrix elements free from infrared singularities, at the leading log approximation. This has been obtained by extending from QED [10] to QCD the concept of classical currents associated with the external particles to incorporate two new properties: colour and the appearance of an effective coupling constant $\bar{\alpha}(\tau)$, where τ is the infrared variable. The inclusive cross sections obtained at this level still develop mass singularities in the limit $m \to 0$, where m is the fermion mass. According to the Kinoshita-Lee-Nauenberg theorem [9], these mass singularities have to be cancelled in this limit in the observable cross sections, through the hard emission of quanta collinear with the external particles. The extension of the coherent-state formalism to include this hard component at all orders finally led to a definition of a super-inclusive or jet cross section [7] which is finite and may be confronted with experiment.

In the present paper we extend the work of ref. [7] in many aspects. By a more accurate kinematical analysis of the emission of hard quanta, we are able to include those finite ϵ corrections which are proportional to ln δ (see below for definitions) for quark jets.

Using the Altarelli-Parisi [11] quark and gluon probabilities we extend our formulae to include gluon jets, in agreement with lowest-order calculations [12]. Finite ϵ terms are explicitly taken into account. A general formula for an arbitrary number of quark and gluon jets is then obtained.

The transverse-momentum properties of the jets are further investigated. In particular we propose new criteria to test the QCD jet predictions in e^+e^- annihilation based on properties which rest on the exponentiated forms obtained upon summation of all orders in perturbation theory. More explicitly, by taking into account the conservation of transverse momentum, we consider the total K_T distribution of the jets which can be directly measured in the actual experiments. We predict the Q^2 dependence of $\langle K_T^2 \rangle$ for quark and gluon jets. We explicitly confirm the general findings [12,13] that gluon jets are broader than quark jets.

Finally, we present a phenomenological extrapolation of our formalism to a K_T region not directly accessible to perturbation theory and compare our predictions with the transverse momentum distributions of single hadrons measured at SPEAR [1]. Our analysis shows good agreement with data at various energies. Predictions are also made for energies accessible in the near feature with PETRA and PEP machines.

A short presentation of our results has been given in ref. [14].

2. Jet cross sections

Our analysis begins with a formula for a jet cross section, derived in ref. [7], for

$e^+e^- \to q\bar{q}$:

$$d\sigma_{\rm super} = d\sigma_0 \exp\left\{-\frac{1}{\pi^2}\int_{\Delta\omega/E}^{1} dx\, \mathcal{P}_{\rm gq}(x) \int_{k_{1\rm T}}^{k_{2\rm T}} \frac{d^2k_{\rm T}}{k_{\rm T}^2}\,\bar{\alpha}(k_{\rm T})\right\}\,, \tag{1}$$

where the gluon distribution due to the quarks is [11]

$$\mathcal{P}_{\rm gq}(x) = C_{\rm F}\left\{\frac{1+(1-x)^2}{x}\right\}, \tag{2}$$

$C_{\rm F} = (N_{\rm c}^2 - 1)/2\,N_{\rm c}$ for SU($N_{\rm c}$) colour, $\bar{\alpha}(k_{\rm T})$ is the running coupling constant

$$\bar{\alpha}(k_{\rm T}) = \frac{12\pi}{(33-2N_{\rm f})}\,\frac{1}{\ln(k_{\rm T}^2/\Lambda^2)}\,,$$

and $d\sigma_0$ is the zeroth-order point-like cross section. The lower limit $k_{1\rm T}$ in the $k_{\rm T}$ integral in eq. (1) defines the maximum transverse momentum relative to the jet axis, namely, $k_{1\rm T} \equiv E\delta$, where δ is the half-angle of the jet cone *, while $k_{2\rm T} = \frac{1}{2}Q = E$. $(\Delta\omega/E)$ represents the fraction of the quark energy which is emitted outside the jet cone. Then the ratio $(d\sigma_{\rm super}/d\sigma_0)$ in eq. (1) gives the probability of finding a fraction $\epsilon \equiv \Delta\omega/2E$ of the total energy $2E$ outside a pair of opposite directed cones of half-angle δ.

First-order expansion of eq. (1) directly gives the result of Sterman and Weinberg, up to terms of order $\epsilon \ln \delta$. A more accurate kinematical analysis shows that the upper limit in x should be restricted to $1 - \Delta\omega/E = 1 - 2\epsilon$. This gives

$$I_{\rm q}(\epsilon) \equiv \int_{2\epsilon}^{1-2\epsilon} dx\,\mathcal{P}_{\rm gq}(x) = -C_{\rm F}\{2\ln 2\epsilon + \tfrac{3}{2} - 2\epsilon(1-2\epsilon) + O(\epsilon^3)\}\,, \tag{3}$$

which agrees with the earlier calculation of Stevenson [15]. We mention in passing that this modification restores the complete symmetry between the quark and gluon x distributions, as expected to hold generally for massless quarks and gluons. In fact the corresponding quark distribution is [11]

$$\mathcal{P}_{\rm qq}(x) = C_{\rm F}\left\{\frac{1+x^2}{1-x}\right\}, \tag{4}$$

and thus

$$\int_{2\epsilon}^{1-2\epsilon} \mathcal{P}_{\rm qq}(x)\,dx = \int_{2\epsilon}^{1-2\epsilon} \mathcal{P}_{\rm gq}(x)\,dx\,. \tag{5}$$

The generalization of eq. (1) to gluon jets is obtained as follows. The probability

* To conform to the general notation [4,5] we have changed the definition of the angle δ which is a half of that used in ref. [7].

$P_{gq}(x)$ in eq. (2) has to be replaced by [11]

$$\mathcal{P}_{gg}(x) + N_f\, \mathcal{P}_{qg}(x) = N_c\left\{\frac{x}{1-x} + \frac{1-x}{x} + x(1-x)\right\} + N_f\,\frac{x^2+(1-x)^2}{2}\,, \tag{6}$$

corresponding to the sum of the probabilities that a gluon radiates gluons or N_f $(q\bar{q})$ pairs. Just as in eq. (3) we have

$$I_g(\epsilon) \equiv \int_{2\epsilon}^{1-2\epsilon} [\mathcal{P}_{gg}(x) + N_f\, \mathcal{P}_{qg}(x)]\, dx$$

$$= -N_c\left\{2\ln 2\epsilon + \left(\tfrac{11}{6} - \tfrac{1}{3}\frac{N_f}{N_c}\right) - 4\epsilon\left(1 - \frac{N_f}{2N_c}\right)(1-2\epsilon) + O(\epsilon^3)\right\}. \tag{7}$$

The logarithmic and the constant terms in eq. (7) agree with those found by explicit perturbative calculations in refs. [12]. Finite terms in ϵ have not as yet been explicitly calculated in perturbation theory.

Notice that our distributions in ϵ for $I_q(\epsilon)$ and $I_g(\epsilon)$ have been obtained without any resource to singular distributions, unlike in the Altarelli-Parisi procedure [11]. The difference arises due to the different range of integration in x.

Using the above results the super-inclusive cross sections for n_q and n_g quark and gluon jets, respectively, are thus given by

$$d\sigma^{(n_q,n_g)}_{\text{super}} = d\sigma_0 \exp\left\{-\frac{1}{\pi}[n_q I_q(\epsilon) + n_g I_g(\epsilon)] \int_{k_{1T}}^{k_{2T}} \frac{d^2k}{k_T}\, \bar{\alpha}(k_T)\right\}, \tag{8}$$

where, for the sake of simplicity, we have assumed the same kinematical limits k_{1T} and k_{2T} for all jets. The jet cross section (8) gives the leading terms in $\ln\delta$. All non-leading terms in δ are lumped in $d\sigma_0$ and have to be calculated for each process separately in perturbation theory. First-order expansion of eq. (8) agrees with the result of Smilga and Visotsky [5].

3. Transverse-momentum distributions

Now let us discuss the transverse-momentum distributions for $e^+e^- \to q\bar{q}$. A similar analysis can be performed in the more general case. From eq. (8), the factor

$$F_q(\epsilon, K_{1T}) \equiv \exp\left\{-\frac{2}{\pi} I_q(\epsilon) \int_{k_{1T}}^{Q/2} \frac{dk_T}{k_T}\, \bar{\alpha}(k_T)\right\}, \tag{9}$$

gives the probability that a fraction ϵ of the total energy Q is emitted outside the oppositely directed cones of maximum transverse momentum $K_{1T} = E\delta$. We have, therefore,

$$\int_0^{K_{1T}} \left(\frac{dP}{dK_T}\right) dK_T = F(\epsilon, K_{1T}) , \tag{10}$$

where (dP/dK_T) defines the K_T distribution, and for simplicity, we have omitted the quark index q. From eqs. (9) and (10) we have

$$\frac{dP}{dK_T} \approx \frac{2}{\pi} I(\epsilon) \frac{\bar{\alpha}(K_T)}{K_T} \left\{\frac{\bar{\alpha}(\frac{1}{2}Q)}{\bar{\alpha}(K_T)}\right\}^{I(\epsilon)\pi b} , \tag{11}$$

where $b = (33 - 2N_f)/12\pi$ and we have used the asymptotic freedom formula for $\bar{\alpha}$. Thus eq. (11) is valid only for $K_T >> \Lambda$. Of course, for our formula (8) to be applicable, we must also require $K_T << \frac{1}{2}Q$.

So far, any correlation in the transverse momentum has been neglected, since it was assumed that the individual emission was statistically independent. To impose transverse-momentum conservation we introduce a factor

$$L(K_{1T}) = \begin{cases} 1 , & \text{for } \sum_n \boldsymbol{k}_{nT} \leqslant \boldsymbol{K}_{1T} , \\ 0 , & \text{otherwise} , \end{cases}$$

$$= \frac{1}{4\pi^2} \int_0^{K_{1T}} d^2K_T \int_{-\infty}^{\infty} d^2\boldsymbol{x} \exp\{i(\sum_n \boldsymbol{k}_{nT} - \boldsymbol{K}_T)\cdot \boldsymbol{x}\} , \tag{12}$$

in the nth-order expansion of the cross section which includes only the hard radiation, defined in ref. [7]. This leads to

$$d\sigma_{\text{hard}} = \frac{d\sigma_0}{4\pi^2} \int_0^{K_{1T}} d^2K_T \int_{-\infty}^{\infty} d^2\boldsymbol{x}\, e^{-i\boldsymbol{x}\cdot \boldsymbol{K}_T} .$$

$$\times \exp\left\{\frac{1}{\pi^2} I(\epsilon) \int_{m^2}^{Q^2/4} \frac{d^2k_T}{k_T^2} \bar{\alpha}(k_T)\, e^{i\boldsymbol{x}\cdot \boldsymbol{k}_T}\right\} . \tag{13}$$

This has to be multiplied by the inclusive cross sections (virtual + real soft radiation) to obtain the super-inclusive one, which is free of the mass singularities:

$$d\sigma_{\text{super}} = \frac{d\sigma_0}{4\pi^2} \int_0^{K_{1T}} d^2K_T \int_{-\infty}^{\infty} d^2\boldsymbol{x}\, e^{-i\boldsymbol{x}\cdot \boldsymbol{K}_T}$$

$$\times \exp\left\{-\frac{1}{\pi^2} I(\epsilon) \int_0^{Q^2/4} \frac{d^2k_T}{k_T^2} \bar{\alpha}(k_T)[1 - e^{i\boldsymbol{x}\cdot \boldsymbol{k}_T}]\right\} . \tag{14}$$

Thus eq. (14) is our replacement of eq. (8) for the two quark jets. We then obtain

$$\frac{d^2P}{d^2K_T} = \frac{1}{4\pi^2}\int\limits_{-\infty}^{\infty} d^2\boldsymbol{x}\, e^{-i\boldsymbol{x}\cdot\boldsymbol{K}_T} \exp\left\{-\frac{1}{\pi^2}I(\epsilon)\int\limits_0^{Q^2/4}\frac{d^2k_T}{k_T^2}\bar{\alpha}(k_T)[1-e^{i\boldsymbol{x}\cdot\boldsymbol{k}_T}]\right\} \tag{15a}$$

$$= \frac{1}{2\pi}\int\limits_0^{\infty} x\, dx\, J_0(x\cdot K_T)\exp\left\{-\frac{2}{\pi}I(\epsilon)\int\limits_0^{Q/2}\frac{dk_T}{k_T}\bar{\alpha}(k_T)[1-J_0(x\cdot k_T]\right\}, \tag{15b}$$

which is properly normalized, for any fixed ϵ, to

$$\int\limits_0^{Q/2}\left(\frac{d^2P}{d^2K_T}\right) d^2K_T = 1\,. \tag{16}$$

Integrating over K_T up to some K_{1T} we find for the probability of finding total transverse momentum K_{1T} and missing fractional energy ϵ

$$P(\epsilon, K_{1T}) = K_{1T}\int\limits_0^{\infty} dx J_1(x\cdot K_T)\exp\left\{-\frac{2I(\epsilon)}{\pi}\int\limits_0^{Q/2}\frac{dk_T}{k_T}\bar{\alpha}(k_T)[1-J_0(x\cdot k_T)]\right\} \tag{17}$$

This equation replaces the earlier expression for $F(\epsilon, K_{1T})$ given in eq. (9) and is the exact analogue of the fractional probability $f(\epsilon, \delta)$ defined by Sterman and Weinberg.

We postpone to sect. 4 a discussion of the range of validity of our eqs. (15) and (17). For completeness, we now give the expressions of the K_T moments of the distribution (15).

Intergrating eq. (15a) by parts, one simply gets

$$K_j\frac{d^2P}{d^2K_T} = \frac{i}{4\pi^2}\int d^2\boldsymbol{x}\, e^{-i(\boldsymbol{x}\cdot\boldsymbol{K}_T)}[\partial_j h(\boldsymbol{x})]\, e^{-h(\boldsymbol{x})}\,, \tag{18}$$

with

$$h(\boldsymbol{x}) = \frac{1}{\pi^2}I(\epsilon)\int\limits_0^{Q/2}\frac{d^2k_T}{k_T^2}\bar{\alpha}(k_T)[1-e^{i(\boldsymbol{k}_T\cdot\boldsymbol{x})}]\,. \tag{19}$$

Similarly,

$$K_lK_j\frac{d^2P}{d^2k_T} = \frac{1}{4\pi^2}\int d^2\boldsymbol{x}\, e^{-i(\boldsymbol{x}\cdot\boldsymbol{K}_T)}[\partial_l\partial_j h(\boldsymbol{x}) - \partial_j h(\boldsymbol{x})\cdot\partial_l h(\boldsymbol{x})]\, e^{-h(\boldsymbol{x})}\,. \tag{20}$$

Therefore, we have

$$\langle K_jK_l\rangle \equiv \int d^2K_T\,\frac{d^2P}{d^2K_T}\;K_jK_l$$

$$= \int d^2\boldsymbol{x}\,\delta(\boldsymbol{x})[\partial_i\partial_j h(\boldsymbol{x}) - \partial_j h(\boldsymbol{x})\cdot\partial_i h(\boldsymbol{x})]\, e^{-h(\boldsymbol{x})}\,, \tag{21}$$

and finally

$$\langle K_T^2\rangle = \frac{I(\epsilon)}{\pi^2}\int_0^{Q/2} d^2k_T\,\bar{\alpha}(k_T)\,. \tag{22}$$

This result gives a simple connection between the average K_T^2 of a jet and the effective coupling constant $\bar{\alpha}$. Using the asymptotic freedom expression for $\bar{\alpha}$ in the perturbative region, namely $k_T > c\Lambda$, we obtain

$$\langle K_T^2\rangle = \frac{I(\epsilon)}{\pi}\int_0^{c\Lambda} dk_T^2\bar{\alpha}(k_T) + \frac{I(\epsilon)\cdot\Lambda^2}{\pi b}\left\{\mathrm{Ei}\left(\ln\frac{Q^2}{4\Lambda^2}\right) - \mathrm{Ei}(\ln c^2)\right\},$$

which for large Q^2 leads to

$$\langle K_T^2\rangle = \text{const} + \frac{I(\epsilon)}{4\pi}\bar{\alpha}(\tfrac{1}{2}Q)\,Q^2\left[1 + O\left(\frac{1}{\ln Q}\right)\right]. \tag{23}$$

The constant term could only be obtained by a knowledge of $\bar{\alpha}(k_T)$ in the non-perturbative region. With the parametrization discussed in sect. 4 we find it negligible for a wide range of c values ($c \gtrsim 1$).

It is interesting to notice that the Q^2 dependence of $\langle K_T^2\rangle$ is simply obtained from eq. (22) as

$$\frac{d}{dQ^2}\langle K_T^2\rangle = \frac{I(\epsilon)}{4\pi}\bar{\alpha}(\tfrac{1}{2}Q)\,, \tag{24}$$

and offers, in our approach, a very simple prediction of perturbative QCD.

As a last remark we observe that for fixed ϵ and Q^2 we obtain the ratio

$$\frac{\langle K_T^2\rangle_{\text{gjet}}}{\langle K_T^2\rangle_{\text{qjet}}} = \frac{I_g(\epsilon)}{I_q(\epsilon)} \xrightarrow[\epsilon\to 0]{} \frac{2N_c^2}{N_c^2 - 1}\,, \tag{25}$$

which predicts gluon jets broader by about a factor of two than quark jets. Similar results have been obtained in refs. [12] and [13].

4. Discussion and numerical results

Some remarks are in order concerning the validity of the equations derived in sect. 3. As is clear from eqs. (15) and (17) the range of k_T integrals extends over a non-perturbative region, $k_T \lesssim c$, where our leading logarithmic approximation is questionable. The influence of the non-perturbative region is negligible for $K_T \gg c\Lambda$. which is the region of interest for testing QCD. For a qualitative understanding we observe that approximating $J_0(z)$ by $\theta(1-z)$ and $J_1(z)$ by $\delta(1-z)$ in eqs. (15)

and (17) one obtains the naive expressions (9) and (11). Therefore, the effect of the correlations, imposed by momentum conservation, is only local.

However, it affects the actual distributions in the perturbative region in a sensible way, leading to departures from the uncorrelated formula (9) and (11) for $K_T >> c\Lambda$.

In the following, we will assume the above correlated distributions to be valid in the entire range of K_T. The problem therefore arises of extrapolating the effective coupling constant in the non-perturbative region. In sect. 5 we will discuss a phenomenological parametrization of $\bar{\alpha}(k_T)$ for $k_T \lesssim c\Lambda$ suggested by a previous analysis [16] which compares all data for the ratio $R = \sigma(e^+e^- \to \mu^+\mu^-)$ in the entire energy range ($4m_\pi^2 \leqslant s \leqslant 60$ GeV^3) with the QCD predictions. This amounts to parametrizing the effective coupling constant as

$$\bar{\alpha}(k_T) = \begin{cases} \dfrac{1}{2b \ln c}\,, & \text{for } k_T \leqslant c\Lambda\,, \\[2ex] \dfrac{1}{2b \ln(k_T/\Lambda)}\,, & \text{for } k_T > c\Lambda\,, \end{cases} \tag{26}$$

with $c \simeq 4$, as discussed in detail in sect. 5.

With the above caveats, our results are presented in figs. 1 and 2 for a two-quark jet and in figs. 3 and 4 for a two-gluon jet. Let us discuss them in detail.

In figs. 1a, b the K_T distributions dP/dK_T of eq. (15) are plotted for different values of $\Delta\omega/E \equiv 2\epsilon$, and Q/Λ, and compared with the naive expression (11). As is clear, a close agreement between the two distributuons is only found for a limited range of K_T/Λ, where the correlations from momentum conservation are less important. Furthermore, the ϵ dependence is rather important, particularly for small ϵ. The agreement is better for the integrated distributions, plotted in figs. 2a, b. They are also in good agreement with those calculated by Ellis and Petronzio [5]. Our results show explicitly the very small influence of the details of the non-perturbative region for a large range of K_T values.

The corresponding results for two-gluon jet distributions are shown in figs. 3 and 4 for the same values of ϵ and Q/Λ. We have used $N_f = 4$. Negligible differences have been found between the case $N_f = 4$ and $N_f = 6$. From an inspection of figs. 3a, b it is evident that momentum correlations have much larger effects on gluon jet distributions as compared to quark jets. In much larger effects on gluon jet distributions as compared to quark jets. In the integrated K_T probabilities, however, some compensations occur such that there is a closer agreement with the naive formulae (see fig. 4). Furthermore, as anticipated earlier (see eq. (25)), there is much more broadening of the momentum distributions in the case of gluon jets (compare figs. 1 and 2 to figs. 3 and 4).

We conclude this section by stressing the fact that our predictions concern global properties of the jets, independent of the properties of the hadronization of partons, so that an experimental verification of our results should represent a rather direct test

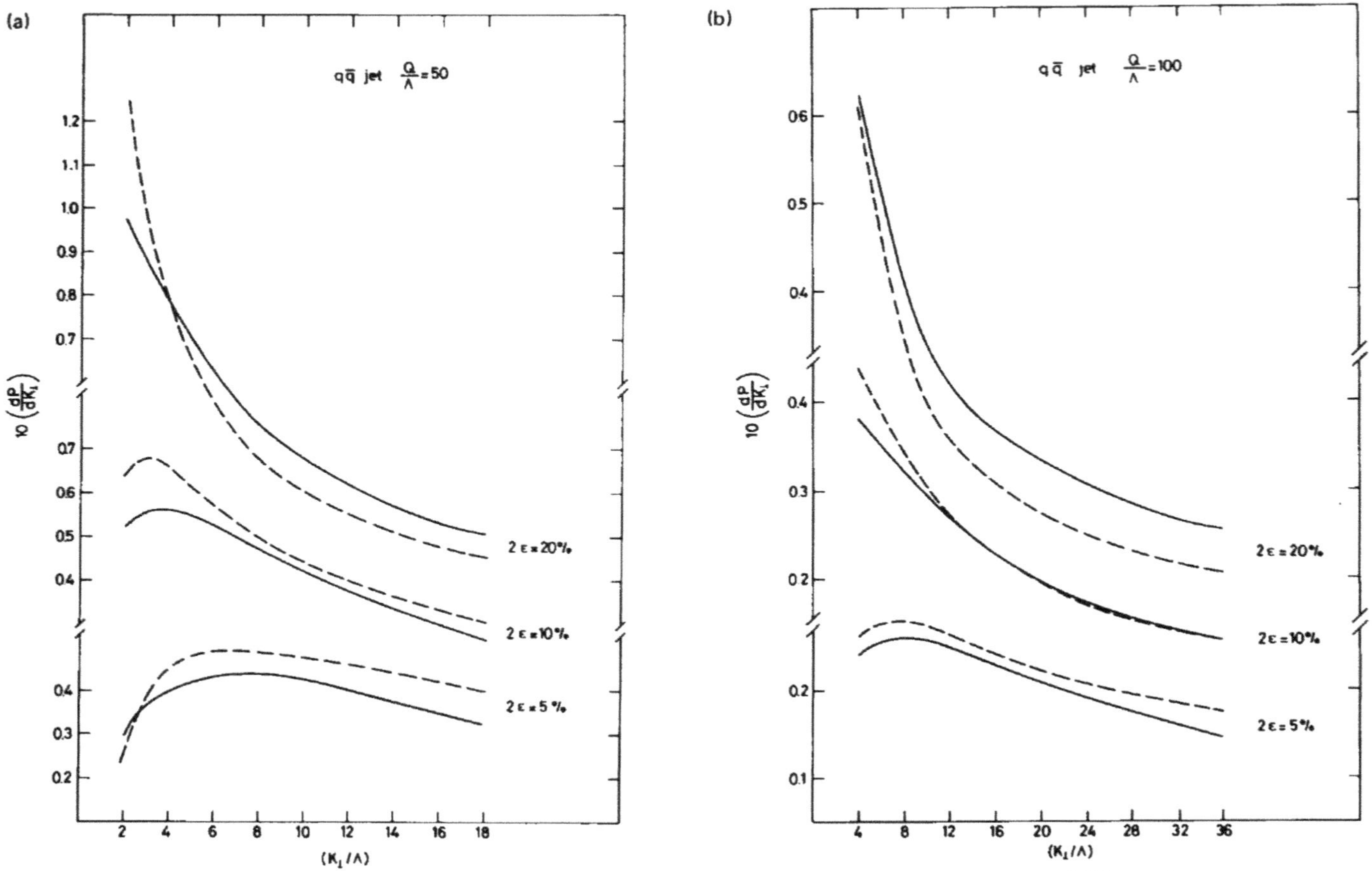

Fig. 1. (a) Total transverse momentum distributions for a $q\bar{q}$ jet at the total energy $Q = 50\ \Lambda$, for various ϵ. The full line refers to the case of transverse momentum conservation taken into account (eqs. (15)). The dashed line refers to the case of uncorrelated emission (eq. (11)). (b) The same, for $Q = 100\ \Lambda$.

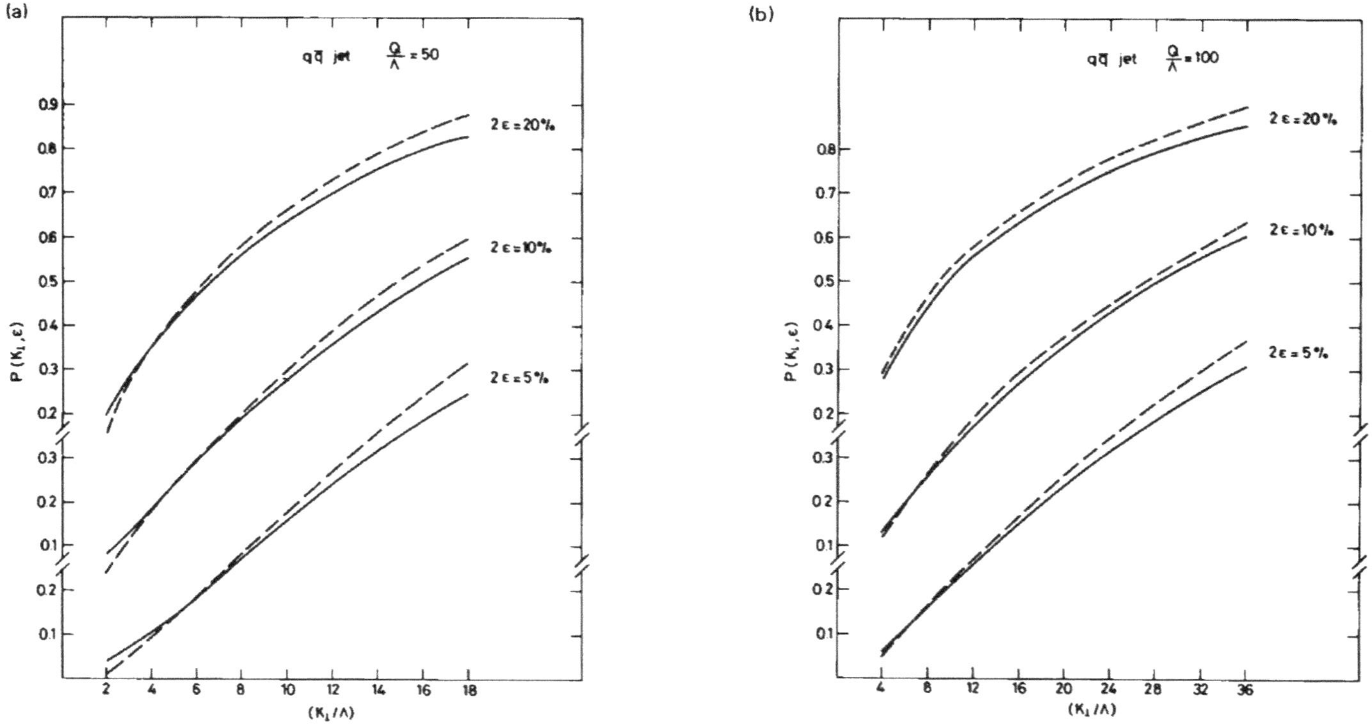

Fig. 2. (a) Probability functions $P(K_T, \epsilon)$ for a $q\bar{q}$ jet at the total energy $Q = 50\ \Lambda$, for various ϵ. Full and dashed lines as in figs. 1 (eqs. (17) and (9), respectively). (b) The same, for $Q = 100\ \Lambda$.

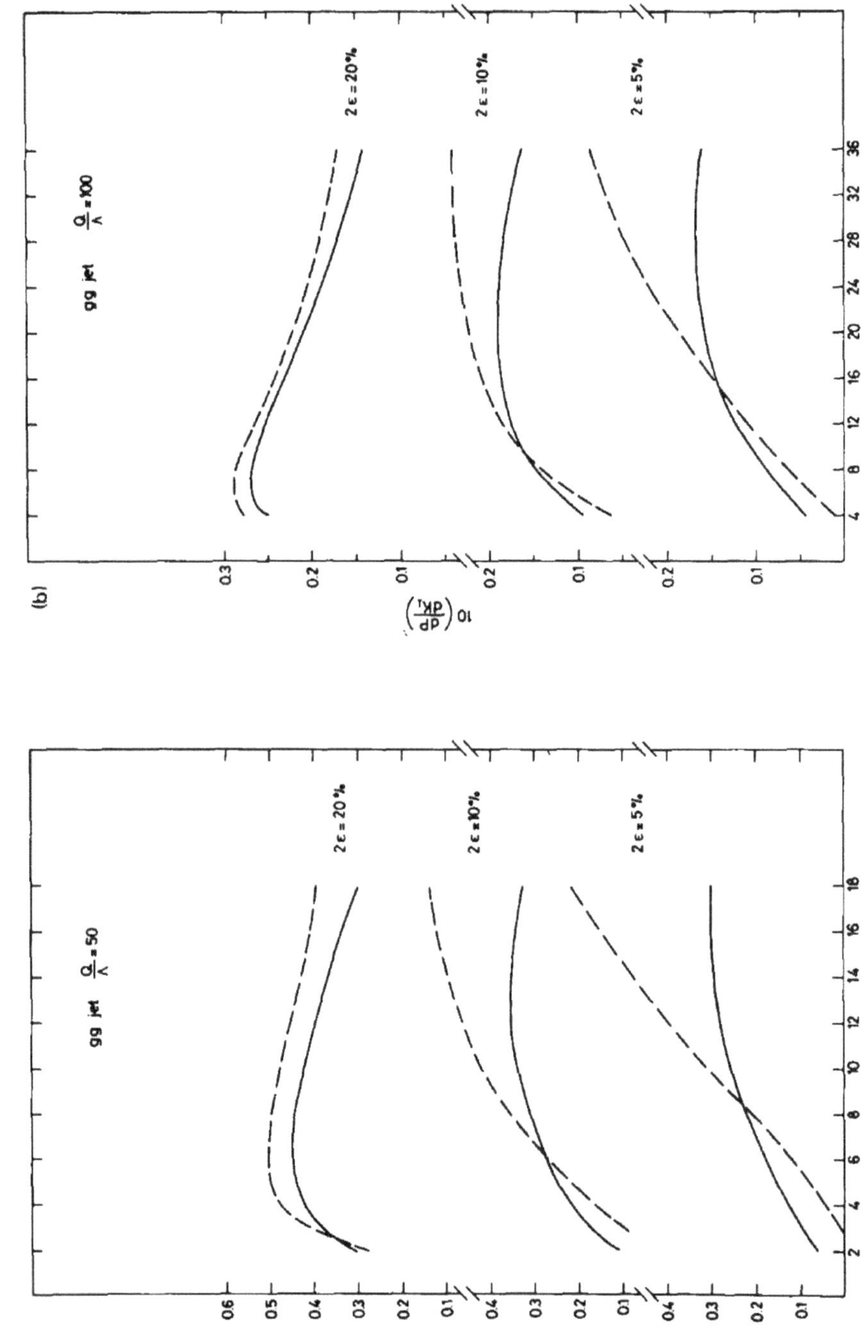

Fig. 3. (a) Total transverse momentum distribution for a gg jet at the total energy $Q = 50\ \Lambda$, for various ϵ. Full and dashed lines as in figs. 1 (eqs. (15) and (11), respectively). (b) The same, for $Q = 100\ \Lambda$.

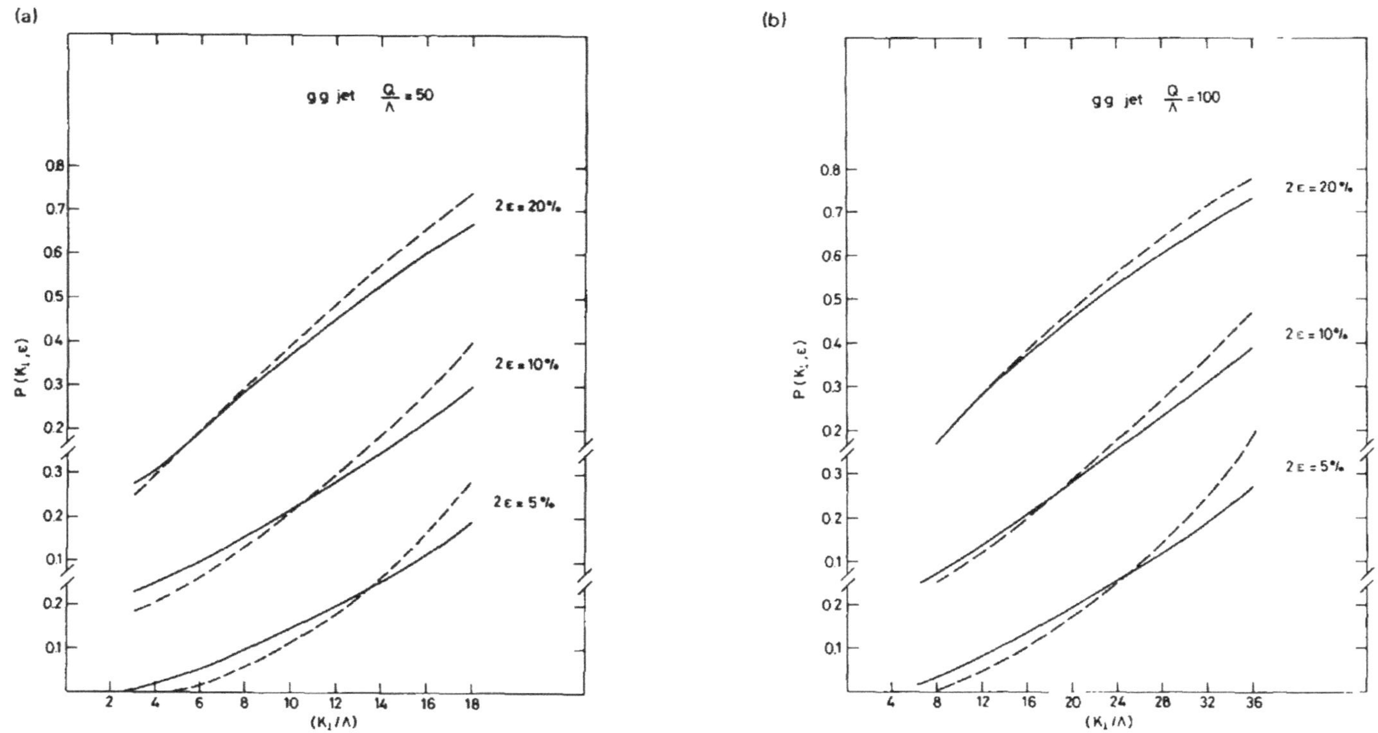

Fig. 4. (a) Probability functions $P(K_T, \epsilon)$ for a gg jet at the total energy $Q = 50\ \Lambda$, for various ϵ. Full and dashed lines as in figs. 1 (eqs. 17) and (9), respectively). (b) The same, for $Q = 100\ \Lambda$.

of QCD. This can be simply obtained by observing the behaviour of the total transverse momentum $\boldsymbol{K}_T$ relative to the jet axis defined as $\boldsymbol{K}_T = \Sigma_i \boldsymbol{k}_{Ti}$ where the sum over i extends to all particles produced in a half-plane orthogonal to the jet axis in a given cone.

5. Phenomenology of hadronic k_T distributions

In this section we discuss the transverse-momentum behaviour of a single hadron, by extrapolating our previous results on K_T distributions in the region $K_T \lesssim c\Lambda$. To achieve this goal, certain extra assumptions have to be made.

First we assume that the above $\mathrm{d}P(K_T)/\mathrm{d}K_T$ can also be used to describe the k_T distribution of an emitted hadron. For this case we limit the maximum transverse momentum allowed to a single hadron at a value $\sim\frac{1}{2}Q\langle n\rangle$, where $\langle n\rangle$ is the average number of particles produced in the e^+e^- annihilation at energy $\sqrt{s} = Q$. This request automatically restricts the K_T range to values $\lesssim c\Lambda$, for present and near-future e^+e^- beam energies. A stated earlier, in this region we use a constant value for $\bar{\alpha}(k_T)$ given by eq. (26). This parametrization is suggested by the observation [16] that, for e^+e^- energies less than about 3 GeV, the average observed value of R,

$$R_{av} \equiv \frac{1}{\bar{s}} \int_{s_0}^{\bar{s}} \mathrm{d}s R_{exp}(s) ,$$

for $\bar{s} \simeq 9\ \mathrm{GeV}^2$ agrees very well with the QCD prediction

$$R(s) = \sum_i Q_i^2 \left(1 + \frac{\alpha(s)}{\pi}\right) ,$$

provided one uses an effective constant value $\alpha(s) = \alpha(\bar{s})$ for $s < \bar{s}$. Using the best fitted value $\Lambda = 0.7$ GeV, which describes all R data for $s > \bar{s}$, and also agrees with a recent analysis [17] of scaling violation in neutrino experiments, we find $c \simeq 4$. Thus for $K_T < c\Lambda$, from eq. (15b) we obtain

$$\frac{\mathrm{d}^2 P}{\mathrm{d}^2 K_T} \simeq \frac{1}{2\pi} \int_0^\infty x\, \mathrm{d}x J_0(x \cdot K_T) \,.$$

$$\times \exp\left\{-\frac{2I(\epsilon)}{\pi} \bar{\alpha}(c\Lambda) \int_0^{Q\langle n\rangle/2} [1 - J_0(x \cdot k_T)] \frac{\mathrm{d}k_T}{k_T}\right\} . \tag{27}$$

Although our final results, discussed below, are obtained by numerical evaluation of eq. (27), we give an approximate expression for eq. (27) which agrees rather well with the exact formula and could be rather useful for future phenomenological use.

With the approximation

$$\int_0^a \frac{dk_T}{k_T}[1 - J_0(xk_T)] \simeq \tfrac{1}{2}\ln(1 + \tfrac{1}{4}a^2x^2)\,, \tag{28}$$

which is obtained by interpolating between the small and large x behaviours of the left-hand side of eq. (28), we finally get

$$\frac{dP}{dk_T} \simeq \frac{2}{a}\frac{\beta}{\Gamma(1+\frac{1}{2}\beta)}\left(\frac{k_T}{a}\right)^{\beta/2} K_{1-\beta/2}\left(\frac{2k_T}{a}\right), \tag{29}$$

where $a = \frac{1}{2}Q\langle n\rangle$, $\beta = 2\bar{\alpha}(c\Lambda)\, I(\epsilon)/\pi$, and $K_{1-\beta/2}(x)$ is the modified Bessel function. In QED this result coincides with that directly obtained from the Bloch-Nordsieck theorem [18].

The distribution (29) exhibits the following limits. For $k_T \ll a$ one finds the power-law behaviour

$$\frac{dP}{dk_T} \simeq \begin{cases} \dfrac{\beta}{a}\dfrac{\Gamma(1-\frac{1}{2}\beta)}{\Gamma(1+\frac{1}{2}\beta)}\left(\dfrac{k_T}{a}\right)^{\beta-1}, & \text{for } \beta < 2\,, \\[2ex] \dfrac{4k_T}{a^2(\beta-2)}\,, & \text{for } \beta > 2\,. \end{cases} \tag{30}$$

On the other hand, in the limit $k_T \gg a$ we obtain

$$\frac{dP}{dk_T} \simeq \frac{\sqrt{\pi}}{a}\frac{\beta}{\Gamma(1+\frac{1}{2}\beta)}\left(\frac{k_T}{a}\right)^{\beta/2-1/2} e^{-(2k_T/a)}\,, \tag{31}$$

which provides an exponential cut-off to the large transverse momenta, independent of $\bar{\alpha}(c\Lambda)$ and ϵ.

Furthermore, the various moments of the distribution (29) are

$$\int \left(\frac{dP}{dk_T}\right) dk_T = 1\,, \tag{32}$$

$$\langle k_T\rangle = \int k_T\left(\frac{dP}{dk_T}\right) dk_T = \tfrac{1}{2}\sqrt{\pi}\tfrac{1}{2}\beta a\,\frac{\Gamma(\frac{1}{2}+\frac{1}{2}\beta)}{\Gamma(1+\frac{1}{2}\beta)}\,, \tag{33}$$

$$\langle k_T^2\rangle = \int k_T^2\left(\frac{dP}{dk_T}\right) dk_T = \tfrac{1}{2}\beta a^2\,, \tag{34}$$

which coincide with those calculated from the exact distribution (27), as can be seen directly, by comparison with eqs. (16) and (22) for $a = \frac{1}{2}Q$ and $\bar{\alpha}(k_T) = \text{const} = \bar{\alpha}(c\Lambda)$. This is a check of the accuracy of our approximate form (29).

Our predictions are compared with the experimental [1]

$$\frac{1}{\sigma}(d\sigma/dk_T)\,,$$

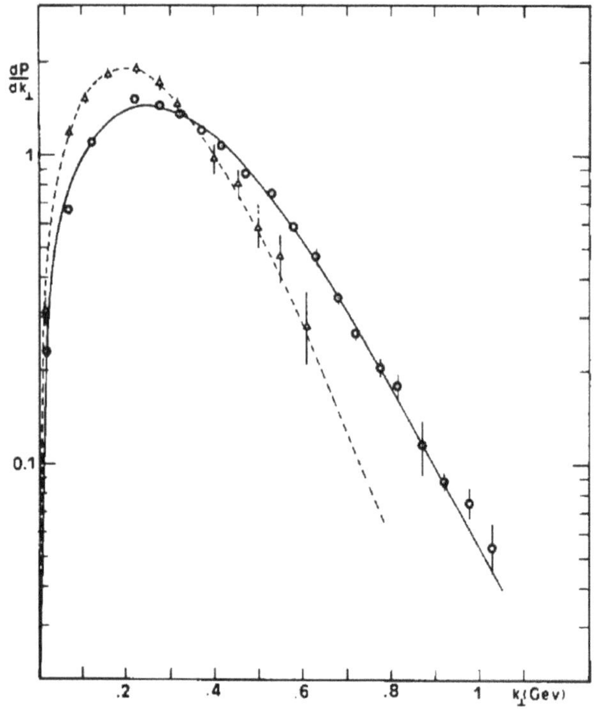

Fig. 5. Single inclusive transverse momentum distributions in $e^+e^- \to q\bar{q} \to$ hadrons at $Q = 3$ and 7.5 GeV. The experimental data are taken from Hanson [1]. The curves are the theoretical predictions (eq. (27)) for $2\epsilon = 0.025$.

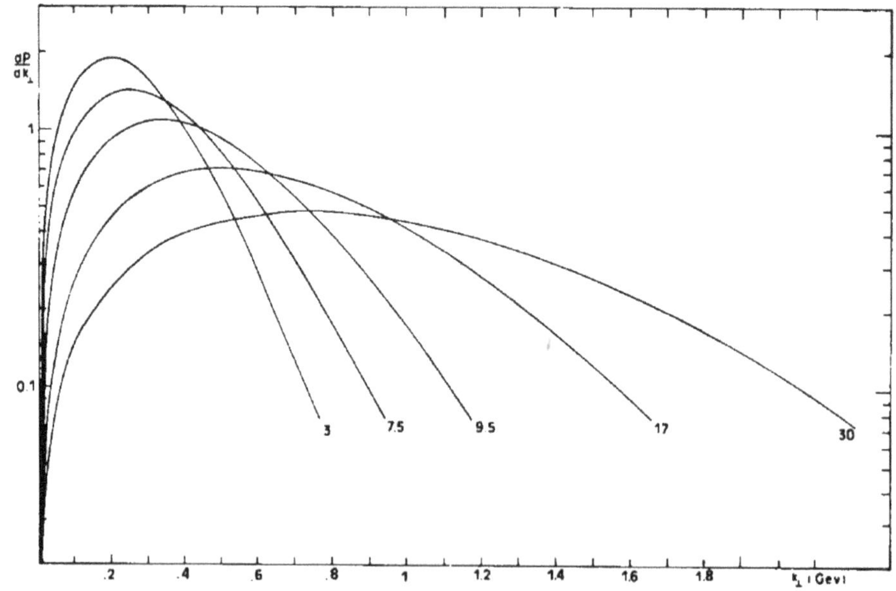

Fig. 6. The same as in fig. 5, for various c.m. total energies, and the same ϵ.

suitably normalized, in fig. 5. Also shown are the results for $Q \simeq 3$ GeV, with the same value of $I(\epsilon)$ and using $\langle n \rangle \sim 6$. The agreement is rather good, both in shape and the Q dependence. We also shown in fig. 6 our predictions for the k_T inclusive distributions at various energies accessible at PETRA and PEP for the same value $2\epsilon = 0.025$ and estimating $\langle n \rangle \sim 2\langle n \rangle_{ch} \sim 2(2.1 + 0.7 \ln s)$. In making comparisons with our formula it is important for fix ϵ for a given experiment. This can be done most straightforwardly by fixing $\langle k_T \rangle$ from the data (as eq. (33)). The effect of the braodening of the distributions (for the same ϵ) is quite evident. If confirmed, it should be considered as a decisive proof of the smoothness of extrapolating perturbative QCD in the infrared region.

Needless to say, a formula analogous to (27) for hadrons produced from gluon jets, for example in the decay $\Upsilon \to 3g$, is trivially obtained with the substitution $n_q I_q(\epsilon) \to n_g I_g(\epsilon)$, and the appropriate kinematical limits.

6. Conclusions

In this work we have analyzed e^+e^- jets in great detail, in the coherent-state formalism, developed for QCD. We have provided explicit expressions for general jet cross sections, which include infinite gluon and (massless) quark-antiquark excitations for each jet. Our approximation also includes finite terms in the energy loss ϵ, in addition to the usual $\ln \delta$ associated with the $\ln \epsilon$ terms. We have checked our results with explicit perturbative calculations, whenever, available.

Explicit expressions for the total K_T distribution, which takes into account the correlations induced by the momentum conservation in a jet, are presented. Predictions are also made for the Q^2 dependence of the mean K_T^2 for quark and gluon jets.

Our discussion has been motivated by the desire to deal with quantities which are accessible experimentally. We hope that such measurements will be forthcoming at PEP, PETRA and LEP energies, thus providing detailed tests of these ideas in QCD.

Finally, we have presented a phenomenological discussion of the single inclusive hadron k_T distributions measured at SPEAR ($W = 3$–7.8 GeV). Our results seem to indicate that perturbation theory can be extrapolated smoothly also in the small k_T region as also found for the total cross sections.

We are grateful to G. Parisi, R. Petronzio and G. Veneziano for useful discussions.

References

[1] G. Hanson, Proc. 13th Rencontre de Moriond (1978), ed. J. Tran Thanh Van (Editions Frontières, France);

B.H. Wiik and G. Wolf, A review of e^+e^- interactions, DESY-78/23 (1978);
PLUTO Collaboration, A study of jets in e^+e^- annihilation into hadrons in the energy range 3.1 to 9.5 GeV, DESY 78/39 (1978);
G. Fontaine, Situation expérimentale de la physique des jets en 1978, 10ème Ecole de Physique des Particules, Gif-sur-Yvette, 4–9 September, 1978, LPC 78-11 (1978).
[2] P.C. Bosetti et al., Nucl. Phys. B142 (1978) 1;
J.G.H. de Groot et al., Phys. Lett. 82B (1979) 292.
[3] V.N. Gribov and L.N. Lipatov, Sov. J. Nucl. Phys. 15 (1972) 438;
H.D. Politzer, Phys. Lett. 70B (1977) 430; Nucl. Phys. B129 (1977) 301;
C.T. Sachrajda, Phys. Lett. 73B (1978) 185; 76B (1978) 100;
D. Amati, R. Pentronzio and G. Veneziano, Nucl. Phys. B140 (1978) 54; B146 (1978) 29;
C.H. Llewllyn Smith, Schladming Lectures 1978, Oxford preprint 47/78 (1978);
Yu.L. Dokshitser, D.I. D'Yakonov and S.I. Troyan, Leningrad Lectures, SLAC report, SLAC-Trans-183 (1978);
R.K. Ellis, H. Georgi, M. Machachek, H.D. Politzer and G. Ross, Phys. Lett. 78B (1978) 281;
W.R. Frazer and J.F. Gunion, Phys. Rev. D19 (1979) 2447.
[4] G. Sterman and S. Weinberg, Phys. Rev. Lett. 39 (1977) 1436.
[5] J. Ellis, M.K. Gaillard and G.G. Ross, Nucl. Phys. B111 (1976) 253;
T.A. De Grand, Y.J. Ng and S.H.H. Tye, Phys. Rev. D16 (1977) 3251;
C.L. Basham, L.S. Brown, S.D. Ellis and S.T. Love, Phys. Rev. D17 (1978) 2298; Phys. Rev. Lett. 41 (1978) 1585;
A. De Rújula, J. Ellis, E.G. Floratos and M.K Gaillard, Nucl. Phys. B138 (1978) 387;
S. Brodsky, T.A. Degrand, R.R. Horgan and D.G. Coyne, Phys. Lett. 73B (1978) 203;
K. Koller and T. Walsh, Nucl. Phys. B140 (1978) 449;
H. Fritzsch and K.H. Streng, Phys. Lett. 74B (1978) 90;
K. Konishi, A. Ukawa and G. Veneziano, Phys. Lett. 78B (1978) 243; 80B (1979) 259;
A.H. Mueller, Phys. Rev. D18 (1978) 3705;
G. Curci and M. Greco, Phys. Lett. 79B (1978) 406;
K. Shizuya, and S.-H. Tye, Phys. Rev. Lett. 41 (1978) 787;
M.B. Einhorn and B.G. Weeks, Nucl. Phys. B146 (1978) 445;
R.K. Ellis and R. Petronzio, Phys. Lett. 80B (1979) 249;
A. Smilga and M. Vysotsky, Nucl. Phys. B150 (1979) 173;
P.M. Stevenson, Phys. Lett. 78B (1978) 451;
A. Ali, J. Körner, Z. Kunst, Z. Willrodt, G. Kramer, G. Schierholz and E. Pietarinen, Phys. Lett. 82B (1979) 285.
[6] M. Greco, F. Palumbo, G. Pancheri-Srivastava and Y. Srivastava, Phys. Lett. 77B (1978) 282.
[7] G. Curci and M. Greco, ref. [5].
[8] F. Bloch and A. Nordsieck, Phys. Rev. 52 (1937) 54.
[9] T. Kinoshita, J. Math. Phys. 3 (1962) 650;
T.D. Lee and M. Nauenberg, Phys. Rev. 133 (1964) 1549.
[10] V. Chung, Phys. Rev. B140 (1965) 1110;
M. Greco and G. Rossi, Nuovo Cim. 50 (1967) 168;
T. Kibble, J. Math. Phys. 9 (1968) 315; Phys. Rev. 173 (1968) 1527; 174 (1968) 1882; 175 (1968) 1624;
N. Papanicolau, Phys. Reports 24 (1976) 229.
[11] G. Altarelli and G. Parisi, Nucl. Phys. B126 (1977) 298.
[12] K. Shizuya and S.-H. Tye; A. Smilga and M. Vysotsky; R.K. Ellis and R. Petronzio, see ref. [5].
[13] K. Konishi, A. Ukawa and G. Veneziano, see ref. [5].

[14] G. Curci, M. Greco and Y. Srivastava, Coherent quark-gluon jets, Preprint NUB-2386, Northeastern University, Boston, February (1979), Phys. Rev. Lett., to appear.
[15] P.M. Stevenson, sec ref. [5];
B.G. Weeks, Corrections to the Sterman–Weinberg jet formula, preprint UMHE 78/49 (1978).
[16] M. Greco, G. Penso and Y. Srivastava, QCD and duality in e^+e^- annihilation, Preprint LNF-78/49 (P) (1978).
[17] R. Barbieri. L. Caneschi, G. Curci and E. d'Emilio, Phys. Lett. 81B (1979) 207.
[18] G. Pancheri-Srivastava and Y. Srivastava, Phys. Rev. D15 (1977) 2915; preprint LNF-78/46 (P) (1978), Phys. Rev. Lett., to appear.

Volume 100B, number 4 PHYSICS LETTERS 9 April 1981

OBSERVATION OF QCD EFFECTS IN TRANSVERSE MOMENTA OF e^+e^- JETS

PLUTO Collaboration

Ch. BERGER, H. GENZEL, R. GRIGULL, W. LACKAS and F. RAUPACH
I. Physikalisches Institut der RWTH Aachen [1], *Germany*

A. KLOVNING, E. LILLESTÖL and J.A. SKARD
University of Bergen [2], *Norway*

H. ACKERMANN, J. BÜRGER, L. CRIEGEE, H.C. DEHNE, A. ESKREYS [3], G. FRANKE, W. GABRIEL [4], Ch. GERKE, G. KNIES, E. LEHMANN, H.D. MERTIENS, U. MICHELSEN, K.H. PAPE, H.D. REICH, M. SCARR [5], B. STELLA [6], U. TIMM, W. WAGNER, G.G. WINTER, W. ZIMMERMANN and M. GRECO [7]
Deutsches Elektronen-Synchrotron DESY, Hamburg, Germany

O. ACHTERBERG, V. BLOBEL [8], L. BOESTEN, V. HEPP [9], H. KAPITZA, B. KOPPITZ, B. LEWENDEL, W. LÜHRSEN, R. van STAA and H. SPITZER
II. Institut für Experimentalphysik der Universität Hamburg [1], *Germany*

C.Y. CHANG, R.G. GLASSER, R.G. KELLOGG, K.H. LAU, R.O. POLVADO, B. SECHI-ZORN, A. SKUJA, G. WELCH and G.T. ZORN
University of Maryland [10], *College Park, USA*

A. BÄCKER [11], F. BARREIRO, S. BRANDT, K. DERIKUM, C. GRUPEN, H.J. MEYER, B. NEUMANN, M. ROST and G. ZECH
Gesamthochschule Siegen [1], *Germany*

H.J. DAUM, H. MEYER, O. MEYER, M. RÖSSLER and D. SCHMIDT
Gesamthochschule Wuppertal [1], *Germany*

Received 20 November 1980

Probability distributions of the total transverse momenta ($K_\perp$) of charged particles produced in hadronic jets in e^+e^- annihilations have been measured for center of mass energies in the range from 9.2 to 31.6 GeV. A linear increase of the average $K_\perp^2$ with Q^2 is observed. The data are successfully compared with high order QCD predictions (according to a simple $q\bar{q}$ picture supplemented by multiple emission of soft gluons). Deviations from this picture at the highest energies and large $K_\perp$ are then analyzed in terms of hard gluon bremsstrahlung and qualitative agreement is found with first order QCD predictions. Scaling "in the mean" is found to be valid both for jet and single particle transverse momenta.

[1] Supported by the BMFT, Germany.
[2] Partially supported by the Norwegian Research Council for Science and Humanities.
[3] Now at the Institute of Nuclear Physics, Krakow, Poland.
[4] Now at Max-Planck Institut für Limnologie, Eutin, Germany.
[5] On leave from University of Glasgow, Glasgow, Scotland
[6] On leave from University of Rome, Italy; partially supported by INFN, Sezione di Roma.
[7] Permanent address: Laboratori Nazionali INFN, Frascati, Italy.
[8] Now at CERN, Geneva, Switzerland.
[9] On leave from Heidelberg University, Germany.
[10] Partially supported by Department of Energy, USA.
[11] Now at Harvard University, Cambridge, MA, USA.

0 031–9163/81/0000–0000/$ 02.50 © North-Holland Publishing Company

Hadron production in e^+e^- annihilation, above ≈5 GeV c.m. energy, is well described by the production of a quark–antiquark pair and its subsequent fragmentation into two hadron jets [1,2]. Quantum chromodynamics (QCD), the candidate theory of strong interactions, predicts that the pure $q\bar{q}$ topology will be altered significantly with increasing energy due to gluon emission. Deviations from the two-jet topology have in fact been observed at the highest energies of PETRA and have been interpreted as clear QCD effects (hard gluon bremsstrahlung [3], soft gluon effects observed in acollinearity distributions [4]).

The QCD prediction of hard gluon bremsstrahlung with its consequences such as events with planar three-jet configurations, has usually been tested by comparing experimental distributions of jet variables and of inclusive particle data with specific (partly phenomenological) models. These models consider first and occasionally second order QCD predictions for the production of partons and use empirically derived algorithms to describe the fragmentation of quarks and gluons into hadrons.

In order to provide quantitative and model independent tests of QCD at the parton level, the data must be compared to the theoretical predictions in variables which are insensitive to the details of the hadronization process of quarks and gluons. The total transverse jet momentum has been proposed as such a variable whose measured distribution would be a direct test of QCD beyond the leading order [5]. The theoretical prediction rests on exponentiated forms obtained upon summation of all orders in perturbation theory in the leading dilogarithmic approximation and includes the correlation imposed by transverse momentum conservation in a jet. Due to this conservation, the partonic distribution is expected not to be modified appreciably by the fragmentation of quarks and gluons into real hadrons.

In this letter we present measurements of distributions of the full transverse momentum ($K_\perp$) for the charged particles of hadronic jets produced in e^+e^- annihilations for c.m. energies in the range from 9.2 to 31.6 GeV and compare them to those predictions. Our data were taken with the magnetic detector PLUTO at the e^+e^- storage rings DORIS and PETRA. The detector and data reduction have been described elsewhere [6]. The results of the analysis of hadronic final states in terms of two- and three-jets have been published [2,4,6].

In the present analysis we determine for each event the direction of maximum thrust for all particles (charged and neutral). The full transverse momentum $K_\perp$ is defined by the following procedure. We first divide each event by a plane orthogonal to the thrust axis and treat the particles on each side separately (thus two $K_\perp$ entries per event are obtained). We then compute for each side $K_\perp = |\Sigma_i^N \boldsymbol{p}_{\perp i}|$, where $\boldsymbol{p}_{\perp i}$ is the transverse momentum vector of a particle relative to the thrust axis and the sum runs over the particles on one side of an arbitrary dividing plane containing that axis. This dividing plane is initially chosen to contain the beam axis, but in order to avoid dynamical biases the dividing plane is then rotated around the thrust axis by ±120° and $K_\perp$ is measured in each case. The average value of these three measurements is used to obtain the $K_\perp$ distributions.

Since particles travelling close to the beam axis escape the detector acceptance, we reduce the contribution of incompletely measured jets by considering only events with thrust axes inclined more than 45° to the beam axis. Constraints have been applied to the measured quantities to ensure that energy is conserved and that the missing mass is positive. Only charged particles, whose pattern recognition and momentum resolution is better than that of neutrals, have been used for computing the $K_\perp$'s and $p_\perp$'s. $K_\perp$ for charged particles is smaller than or equal to the total $K_\perp$. However, we can overcome this by exploiting a nice feature of the expected distribution, namely the "scaling in the mean" (see below): if we measure the reduced total transverse momentum $K_\perp/\langle K_\perp\rangle$, where $\langle K_\perp\rangle$ is the average observed $K_\perp$, the distributions for all particles or for a sample of them are expected to become identical.

All experimental distributions have been corrected for acceptance, detection efficiency, and resolution of our detector as well as for electromagnetic radiation using a Monte Carlo calculation. The hadron production model is based on that of Hoyer et al. [7] and includes hard gluon emission as well as 4 or 5 quark flavours in the final state (i.e. u, d, s, c quarks for 9.4 GeV data and u, d, s, c, b quarks for 12–31.6 GeV data). For our selected events, the correction is completely negligible at high $K_\perp$ values and amounts to an ≈8% change at the most probable $K_\perp$ at each energy. We have checked that our corrections are practically model independent by comparison to a different Monte Carlo calculation based on a model by Ali et al. [8]. Both these models are partly phenomenological and reproduce the measured distributions quite well. In the final distributions some of the energy points (9.4, 27.5, 30.0, 31.05) are averages over energy intervals, typically of 1 GeV width.

Theoretically the probability distribution $(dP/dK_\perp)$ of the $q\bar{q}$ jet transverse momentum is given by [5,9] [‡1]

$$\frac{dP}{dK_\perp} = K_\perp \int_0^\infty \xi d\xi J_0(\xi K_\perp) \exp\left(-\frac{16}{3\pi} \times \int_0^{\bar{k}_\perp} \frac{dq_\perp}{q_\perp} \ln(Q/q_\perp)\,\alpha_s(q_\perp)[1 - J_0(\xi q_\perp)]\right), \tag{1}$$

where $\alpha_s(q_\perp)$ is the running coupling constant and $\bar{k}_\perp$ is the effective phase space limit for the emitted gluons, which is proportional to Q. (Its actual determination for our analysis will be discussed below.) Eq. (1) has been derived [5] by an explicit sum of all orders in α_s, in the leading double logarithmic approximation and assuming massless quarks. Transverse momentum conservation is also explicitly taken into account. From eq. (1), which is properly normalized to unity, one also gets:

$$\langle K_\perp^2 \rangle = \frac{4}{3\pi} \int_0^{\bar{k}_\perp} d(q_\perp^2) \ln(Q^2/q_\perp^2)\,\alpha_s(q_\perp), \tag{2}$$

which for large Q^2 leads to $\langle K_\perp^2 \rangle \propto Q^2$ + const., up to logarithmic terms.

To extend the distribution (1) to the region of small $K_\perp$, $\alpha_s(k_\perp)$ is replaced by an effective coupling constant $\alpha_{\text{eff}} \approx 0.48$. This value has been determined in a previous independent analysis [11], based on duality, which successfully compares all data for the ratio

$$R \equiv \sigma(e^+e^- \to \text{hadrons})/\sigma(e^+e^- \to \mu^+\mu^-),$$

with first order QCD predictions. Then the low energy domain ($Q \lesssim 3$ GeV) determines an effective value of α_s as given above, whereas at higher energies, the usual asymptotic freedom estimated $\alpha_s(Q) = 1/[2b \ln(Q/\Lambda)]$ is used, with $b = 25/(12\pi)$ and $\Lambda \approx 0.7$ GeV. Then eq. (2) can be rewritten as

$$\langle K_\perp^2 \rangle_{q\bar{q}} = \tfrac{1}{2}\beta \bar{k}_\perp^2, \tag{3}$$

with

$$\beta = \tfrac{8}{3}(\alpha_{\text{eff}}/\pi)[\ln(Q^2/\bar{k}_\perp^2) + 1]. \tag{4}$$

In order to compare the predicted $(dP/dK_\perp)$ with the data at all energies, we shall make use of the observation [12,5] that this distribution is expected to approximately scale with Q^2 as a function of $K_\perp/\langle K_\perp \rangle$, where the average $\langle K_\perp \rangle_{q\bar{q}}$ of (1) is given by

$$\langle K_\perp \rangle_{q\bar{q}} \simeq \tfrac{1}{4}\sqrt{\pi}\,\beta \bar{k}_\perp \frac{\Gamma(1/2 + \beta/2)}{\Gamma(1 + \beta/2)}. \tag{5}$$

Then the distribution $dP/d(K_\perp/\langle K_\perp \rangle)$ depends only on β, and using eqs. (3)–(5) on α_{eff}, i.e. it is expected to be approximately independent of energy ("scaling in the mean").

We first show in fig. 1 (circles) the energy dependence of the experimental $\langle K_\perp^2 \rangle$ of all charged particles. The full line is a linear fit to our results. The almost linear increase of $\langle K_\perp^2 \rangle$ with Q^2 is much more pronounced than for the case of single particle inclusive production [3]. A linear rise is expected from multiple soft gluon radiation described by eq. (2) [see also eq. (3): the slope is not directly predicted]. Notice, however, that hard gluon bremsstrahlung would also produce a similar increase with Q^2 [13]. Therefore the observed feature should result from the two combined effects. The linear fit to the data (cir-

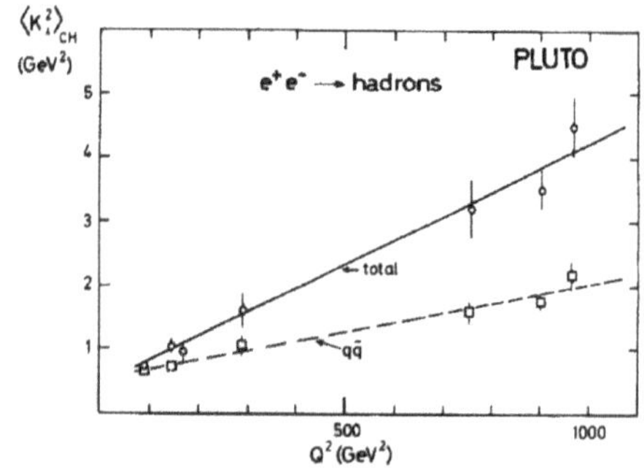

Fig. 1. Average squared transverse momentum relative to the thrust axis (from charged particles only) versus squared c.m. energy. Circles: directly measured values. Squares: fitted $\langle K_\perp^2 \rangle_{\text{ch}}$ attributed to $q\bar{q}$ jets (see text). Both full and dashed lines are linear fits to the data.

[‡1] For the Drell–Yan process the same formula has been proposed by Parisi and Petronzio [10].

cles) gives

$$\langle K_\perp^2\rangle_{\rm ch} = (0.54 \pm 0.01) + (3.84 \pm 0.01)10^{-3}Q^2(\mathrm{GeV}^2),$$

with negligible errors. The square points will be explained below.

Next we show in fig. 2 the experimental distribution $dP/d(K_\perp/\langle K_\perp\rangle)$ at several c.m. energies. The data are compared with the absolute prediction of eq. (1) in the reduced variable $K_\perp/\langle K_\perp\rangle$, computed for $Q = 9.4$ GeV and $\alpha_{\rm eff} = 0.48$ (full curve). In addition to the full transverse momentum data we also plot in fig. 2 the distribution of the single particle reduced transverse momentum $dP/d(p_\perp/\langle p_\perp\rangle)$ at 9.4 and 30.0 GeV. It has been observed [5,12] that in fact the extrapolation of the theoretical $K_\perp$ distributions at very low momenta describes very accurately the $p_\perp$ distributions.

As demonstrated in fig. 2, our results nicely overlap in a wide range of energies and agree also quantitatively with the theoretical prediction. It is rather striking that the same universal curve describes the $p_\perp$ and $K_\perp$ distributions in a wide range of transverse momenta ($p_\perp \lesssim 1.5$ GeV, $K_\perp \lesssim 4$ GeV). This scaling in the mean had been observed empirically for inclusive $p_\perp$ distributions in pp interactions [14], but only recently it was shown to be deducible from QCD [12] ‡2.

A more detailed analysis shows, however, that deviations occur from the simple $q\bar{q}$ picture at high energies and large $K_\perp$, as is expected if hard gluon bremsstrahlung sets in. We therefore make a separate analysis of the $K_\perp$ distributions at each energy, assuming a mixture of $q\bar{q}$ jets [eq. (1)] and $q\bar{q}g$ jets. For the latter, the perturbative first order $K_\perp$ distribution with respect to the thrust axis is given by

$$\frac{1}{\sigma_0}\frac{d\sigma(q\bar{q}g)}{dx_\perp^2} = \frac{1}{\sigma_0}\int_{T_{\rm min}}^{1-x_\perp^2}\frac{d\sigma(q\bar{q}g)}{dT\,dx_\perp^2}\,dT, \qquad (6)$$

where

$$x_\perp^2 = (4K_\perp^2/Q^2) = (4/T^2)(1-x_q)(1-x_{\bar{q}})(1-x_g),$$

T is the thrust variable, σ_0 is the lowest order $q\bar{q}$ cross section, $T_{\rm min}$ is the solution of

$$x_\perp^2 = (4/T_{\rm min}^2)(1-T_{\rm min})^2(2T_{\rm min})^{-1},$$

and the integrand of the rhs of eq. (6) is well known [13].

In figs. 3 and 4 we present our probability distributions for $K_\perp$ at various energies and compare them with the theoretical predictions. The full curves describing $q\bar{q}$ jets have been obtained from eq. (1) for $\alpha_{\rm eff} = 0.48$,

‡2 The analogy between e^+e^- and pp jets is presently being exploited at CERN by the CERN–Bologna–Frascati–Bari Collaboration (see for instance Basile et al. [15]).

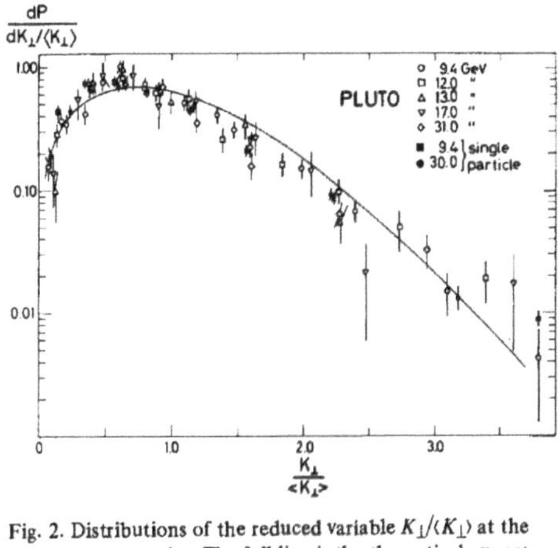

Fig. 2. Distributions of the reduced variable $K_\perp/\langle K_\perp\rangle$ at the quoted c.m. energies. The full line is the theoretical expectation computed at 9.4 GeV. Also shown (full points) are the experimental distributions of the corresponding single particle reduced variable $p_\perp/\langle p_\perp\rangle$ at 9.4 and 30 GeV. (In order not to overload the figure, we show here only one of our high energy distributions, which are reported in fig. 4.)

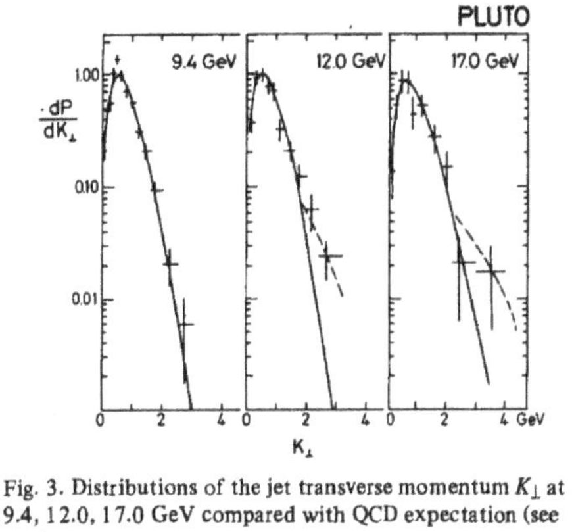

Fig. 3. Distributions of the jet transverse momentum $K_\perp$ at 9.4, 12.0, 17.0 GeV compared with QCD expectation (see text).

Volume 100B, number 4 PHYSICS LETTERS 9 April 1981

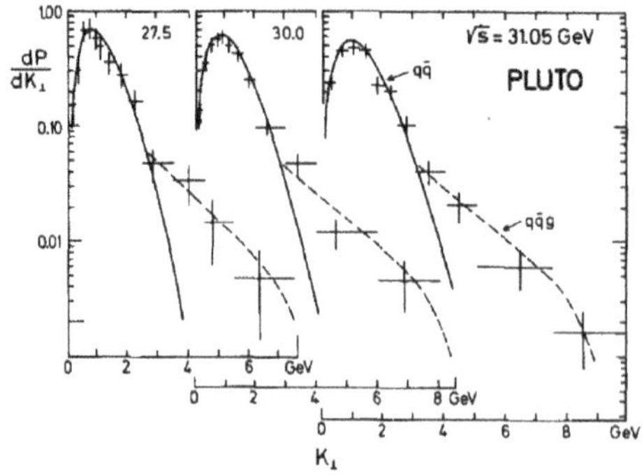

Fig. 4. Same as fig. 3 for 27.5, 30.0 and 31.05 GeV average c.m. energies. (The two last points for 30 GeV do not appear in fig. 2, being out of range.)

fitting $\langle K_\perp^2 \rangle_{q\bar{q}}$ at each energy to all but the highest $K_\perp$ points. [This, according to eq. (3), is equivalent to determining $\bar{k}_\perp^2$.] The resulting values for $\langle K_\perp^2 \rangle_{q\bar{q}}$ are plotted in fig. 1 (squares, broken line). We get a linear increase also in this case as expected, and it is clear that the broadening of a QCD q$\bar{\text{q}}$ jet with Q^2 accounts only for about a half of the actual increase of the measured $\langle K_\perp^2 \rangle_{ch}$.

The large $K_\perp$ tail in figs. 3, 4 (dashed curve) is obtained from eq. (6), which also scales with $\langle K_\perp \rangle$. Because only a fraction of the gluon's transverse momentum is retained in our averaging procedure, an appropriate rescaling factor has to be applied. This has been fixed by normalizing the full (soft + hard) $K_\perp$ distribution to the data. The stability and uniqueness of our procedure has been carefully checked. A value $\Lambda \simeq 0.7$ has been used for $\alpha_s(Q)$. Our results do not depend appreciably on $\alpha_s(Q)$ due to the lack of absolute normalization.

In fig. 3 we observe that the 9.4 GeV data are very well described by simply a q$\bar{\text{q}}$ jet, the hard gluon contribution being small. As Q increases the effect of hard gluon bremsstrahlung becomes more and more important, being essential to understand the data at large $K_\perp$ at the highest values of Q (fig. 4). It also accounts for the difference in the two lines in fig. 1.

We conclude that the $K_\perp$ distributions as well as the first observation of the linear increase of $\langle K_\perp^2 \rangle_{ch}$ with Q^2 give strong support to soft gluon resumming formulae in QCD and to the process of single hard gluon bremsstrahlung. We note that our analysis procedure makes no use of phenomenological models of the hadronization of quarks and gluons. We also find that scaling in the mean is valid for both the $K_\perp$ as well as the inclusive $p_\perp$ distributions (it is expected to set in more rapidly for $K_\perp$ than $p_\perp$ as a function of Q), perhaps indicating that hadronization retains in transverse momentum much of the primary parton properties. Moreover, a universal scaling curve approximately describes the relevant distributions at all energies suggesting a similar basic mechanism for different processes (e^+e^- and pp). The agreement of the data and the predictions also support the reliability of the parametrization of QCD at small Q^2 with an effective coupling constant.

We would like to acknowledge the stimulating support of Y. Srivastava in the early stage of this analysis. One of us (M. Greco) gratefully acknowledges the use of computer time at his home institution. We wish to thank Professors H. Schopper, G. Voss, E. Lohrmann and Dr. G. Söhngen for their valuable support. We are indebted to the PETRA machine group and the DESY computer center for their excellent performance during the experiment. We gratefully acknowledge the efforts of all engineers and technicians of the collaborating institutions who have participated in the construction and the maintenance of the apparatus.

References

[1] See for instance, the review by R. Devenish, in: Surveys in high energy physics, Vol. 1 (1979) p. 1.
[2] PLUTO Collab., Ch. Berger et al., Phys. Lett. 78B (1978) 176.
[3] TASSO Collab., R. Brandelik et al., Phys. Lett. 86B (1979) 243; 94B (1980) 437;
PLUTO Collab., Ch. Berger et al., Phys. Lett. 86B (1979) 418;
MARK J Collab., D.P. Barber et al., Phys. Rev. Lett. 43 (1979) 830; Phys. Lett. 89B (1979) 139;
JADE Collab., W. Bartel et al., Phys. Lett. 91B (1980) 142;
A. Ali, E. Pietarinen, G. Kramer and J. Willrodt, Phys. Lett. 93B (1980) 155.
[4] PLUTO Collab., Ch. Berger et al., Phys. Lett. 90B (1980) 312.
[5] G. Curci, M. Greco and Y. Srivastava, Phys. Rev. Lett. 43 (1979) 834; Nucl. Phys. B159 (1979) 451.
[6] PLUTO Collab., Ch. Berger et al., Phys. Lett. 76B (1978) 243; 66B (1977) 395; 81B (1979) 410.

[7] P. Hoyer et al., Nucl. Phys. 161B (1979) 349.
[8] A. Ali, E. Pietarinen, G. Kramer and J. Willrodt, DESY 79/86 (1979).
[9] M. Greco, Proc. XVe Rencontre de Moriond (1980), ed. J. Tran Thanh Van.
[10] G. Parisi and R. Petronzio, Nucl. Phys. B154 (1979) 427.
[11] M. Greco, Y. Srivastava and G. Penso, Phys. Rev. D21 (1980) 2520.
[12] G. Pancheri-Srivastava and Y. Srivastava, Phys. Rev. D21 (1980) 95.
[13] T.A. De Grand, Y.J. Ng and S.-H.H. Tje, Phys. Rev. D16 (1977) 3251;
A. de Rújula, J. Ellis, E.G. Floratos and M.K. Gaillard, Nucl. Phys. B138 (1978) 387.
[14] F.T. Dao et al., Phys. Rev. Lett. 33 (1974) 389;
A. Laasanen et al., Phys. Rev. Lett. 38 (1977) 1.
[15] M. Basile et al., Phys. Lett. 92B (1980) 367.

Nuclear Physics B246 (1984) 12–44

VECTOR BOSON PRODUCTION AT COLLIDERS: A THEORETICAL REAPPRAISAL

G. ALTARELLI

CERN, Geneva, Switzerland and Dipartimento di Fisica, Università "La Sapienza", Roma, Italy
INFN, Sezioue di Roma, Italy

R.K. ELLIS*

INFN, Sezione di Roma, Italy

M. GRECO and G. MARTINELLI

INFN, Laboratori Nazionali di Frascati, Italy

Received 2 May 1984
(Revised 13 June 1984)

We study the production of vector bosons in hadron-hadron collisions via the Drell-Yan mechanism in QCD. Our treatment of the transverse momentum and rapidity distributions of the produced bosons takes all available theoretical principles and results into account in a systematic way. The resulting q_T distribution reduces to the perturbative limit for large q_T, includes the summation of soft gluons and reproduces the known results for the total cross section. A full numerical analysis of W and Z cross sections at collider and tevatron energies is made.

1. Introduction

The production of W and Z bosons at the CERN $p\bar{p}$ collider [1,2] tests the Drell-Yan mechanism [3] in a completely new energy regime. The total cross section for vector boson production σ and the rapidity differential cross section $d\sigma/dy$ are predicted by the QCD-improved parton model [4] as an expansion in the strong coupling constant α_s. The corrections of order α_s to these cross sections have been calculated and found to be important [5,6]. They increase the naive parton model prediction by an energy and rapidity dependent factor commonly referred to as the "*K*-factor". At fixed target energies the $O(\alpha_s)$ corrections are dangerously big and resummation techniques must be invoked [7] in an attempt to control the perturbation series. At collider energies their size is reduced because the coupling constant is smaller and, for the production of weak intermediate bosons, they lead to a

* Address after 1 April 1984: FERMILAB, Batavia, IL 60510, USA

correction of about 30%. The total cross section is therefore more reliably predicted by perturbation theory at these energies with a smaller theoretical error on the overall normalization.

The prediction of the boson transverse momentum distribution is more subtle, since all order effects need to be taken into account. Renormalization group-improved perturbation theory is valid when the transverse momentum q_T is of the same order as the vector boson mass Q. The large q_T tail of the transverse momentum distribution was one of the early predictions of the QCD-improved parton model [8, 9]. As q_T becomes less than Q, such that $\Lambda \ll q_T \ll Q$, a new scale is present in the problem and large terms of order

$$\frac{1}{q_T^2}\alpha_s^n(q_T^2)\ln^m(Q^2/q_T^2), \qquad m \leqslant 2n-1,$$

occur, forcing the consideration of all orders in n. These terms are characteristic of a theory with massless vector gluons. Fortunately, in the leading double logarithmic approximation (DLA: $m = 2n - 1$) these terms can be reliably resummed. This resummation was first attempted by Dokshitzer-Diakonov-Troyan (DDT) [10] and subsequently modified and consolidated [11]. A consistent framework for going beyond the leading double logarithmic approximation has been indicated by Collins and Soper [12, 13].

The combination of these results on the q_T distribution with the constraint on the area of the distribution provided by the integrated cross section at $O(\alpha_s)$, allows an essentially complete reconstruction of the q_T distribution to that accuracy. There is some uncertainty due to the parton intrinsic transverse momentum but it is present only in a restricted region at low q_T at collider energies. Despite these theoretical advances and several numerical analyses [14], there is to our knowledge no complete and explicit treatment in the literature, valid both in the region $q_T \sim Q$ and $q_T \ll Q$, which includes all the available information.

In this paper we re-examine the problem of the q_T distribution in Drell-Yan processes. We include in a systematic way the large amount of theoretical information accumulated in recent years. Our final expression for the q_T distribution satisfies the following requirements:

(i) At large q_T, we automatically recover the $O(\alpha_s)$ perturbative distribution coming from one-gluon emission, without the ad hoc introduction of matching procedures between hard and soft radiation.

(ii) In the region $q_T \ll Q$ the soft gluon resummation is performed at leading double logarithmic accuracy. The role of subleading terms in the summation is also discussed and evaluated.

(iii) Only terms corresponding to the emission of soft gluons, for which the exponentiation can be theoretically justified, are resummed. The proposal to exponentiate the whole first-order contribution, originally proposed by Parisi and

Petronzio [11] and more recently adopted by Halzen, Martin and Scott [15], is at variance with the all-orders analysis of Collins and Soper [12].

(iv) The integral of the q_T distribution reproduces the known results for the $O(\alpha_s)$ total cross sections (i.e. including the "K-factor").

(v) The average value of q_T^2 is also identical with the perturbative result at $O(\alpha_s)$.

(vi) All quantities are expressed in terms of precisely defined quark distribution functions at a specified scale, as for example those determined by the deep inelastic structure function F_2 at the scale Q^2 [5].

Our treatment can be augmented with higher-order terms as and when these become available. At present, it is not possible to include $O(\alpha_s^2)$ terms in an entirely consistent way because a complete calculation is lacking. The calculation of ($q\bar{q} \to B + X$) at non-zero q_T exists at order α_s^2 [16] and was found to be numerically important at fixed target energies, but the complete two-loop calculation has not yet been performed. In the literature one can find only one correction to the double logarithmic approximation, first evaluated by Kodaira and Trentadue [17] and recently confirmed [18] using the results of ref. [16]. We include this term as a way of testing the theoretical stability of our results.

We also make a thorough numerical analysis of the W and Z boson cross sections and distributions. We pay special attention to the quantitative uncertainties due to lack of precise knowledge of the input parameters. These parameters are principally Λ_{QCD}, the parton distribution functions, the intrinsic transverse momentum and the scale of the running coupling constant in front of the $O(\alpha_s)$ terms.

Consider a proton and antiproton with momenta P_1 and P_2 respectively, which collide at total centre-of-mass energy $\sqrt{S}$ to produce a vector boson B of momentum q. In the centre-of-mass frame of the incoming hadrons, these momenta have components

$$P_1 = \tfrac{1}{2}\sqrt{S}\,(1;0,0,1),$$

$$P_2 = \tfrac{1}{2}\sqrt{S}\,(1;0,0,-1),$$

$$q = (q_0;\boldsymbol{q}_T,q_3), \qquad q^2 \equiv Q^2, \qquad q_T^2 \equiv \boldsymbol{q}_T^2. \tag{1}$$

From these variables we can define the hadronic invariants

$$T = (P_1 - q)^2, \qquad U = (P_2 - q)^2, \tag{2}$$

and the rapidity y:

$$y = \tfrac{1}{2}\ln\frac{q_0 + q_3}{q_0 - q_3} = \tfrac{1}{2}\ln\frac{Q^2 - T}{Q^2 - U}. \tag{3}$$

According to the parton model, the production of the boson proceeds through the interaction of a parton of momentum p_1 in the proton with a parton of momentum p_2 in the antiproton. The invariant cross section is given by

$$\frac{\mathrm{d}\sigma}{\mathrm{d}^2\boldsymbol{q}_\mathrm{T}\,\mathrm{d}y} = \sum_{i,j}\int \mathrm{d}x_1\,\mathrm{d}x_2\, f_i(x_1) f_j(x_2) \left[\frac{s}{\pi}\frac{\mathrm{d}\sigma_{ij}}{\mathrm{d}t\,\mathrm{d}u}\right]_{\substack{p_1 = x_1 \\ p_2 = x_2 P_2}}, \tag{4}$$

where f_i, f_j are the distributions of partons of type i, j in the parent hadrons and the parton cross sections are expressed in terms of the partonic variables

$$s = (p_1 + p_2)^2, \qquad t = (p_1 - q)^2, \qquad u = (p_2 - q)^2. \tag{5}$$

The lowest-order parton cross section gives a contribution only at $\boldsymbol{q}_\mathrm{T} = 0$:

$$\frac{\mathrm{d}\sigma}{\mathrm{d}^2\boldsymbol{q}_\mathrm{T}\,\mathrm{d}y} = NH(x_1^0, x_2^0)\,\delta^2(\boldsymbol{q}_\mathrm{T}), \tag{6}$$

where N is an overall normalization and H is the product of quark and antiquark parton densities evaluated at the points

$$x_1^0 = \sqrt{\tau}\,\mathrm{e}^{y}, \qquad x_2^0 = \sqrt{\tau}\,\mathrm{e}^{-y}, \qquad \tau = \frac{Q^2}{S}. \tag{7}$$

However, eq. (4), with perturbatively evaluated parton cross sections, is only applicable when there is a single large scale $q_\mathrm{T} \sim Q$. For this region, the cross section is well described by the emission of one [8, 9] or possibly two gluons [16]. As q_T^2 becomes smaller, terms of order

$$\frac{\alpha_\mathrm{s}(q_\mathrm{T}^2)}{\pi}\frac{1}{q_\mathrm{T}^2}\ln\frac{Q^2}{q_\mathrm{T}^2}, \qquad \frac{\alpha_\mathrm{s}(q_\mathrm{T}^2)}{\pi}\frac{1}{q_\mathrm{T}^2}, \tag{8}$$

become large and must be resummed if we want to have a valid perturbative prediction. The divergence at $q_\mathrm{T}^2 = 0$ is only apparent since for $q_\mathrm{T} = 0$, the virtual diagrams intervene and cancel the divergence. We can resum the terms of eq. (8), using all the information which can be extracted from one-parton emission cross sections.

The calculation, which we describe in the next section, proceeds in five steps. None of the steps are mathematically complicated but the length of the formula makes the whole treatment rather heavy. We therefore list the steps in the manipulation of the cross sections so that the reader who is not interested in the mathematical details may skip sect. 2 and proceed to the answer given in sect. 3.

1.1 STEP I

Write the cross sections for the processes

$$\begin{aligned} &q+\bar{q}\to B+g,\\ &q+g\to B+q,\\ &q+\bar{q}\to B, \end{aligned} \tag{9}$$

up to and including terms of order α_s. The cross sections are written in n dimensions so that the singularities present at $q_T^2=0$ are regulated. These cross sections can be found in ref. [5].

1.2. STEP II

Insert these parton cross sections into eq. (4) to derive the corresponding hadronic cross sections. By subtraction and re-addition of the residues of the terms which are singular as $q_T^2\to 0$, we make the poles in $(n-4)$ explicit. The double poles coming from the region of soft emission cancel; the single poles are the collinear singularities which should ultimately be factored into the parton distributions.

1.3. STEP III

Separate the $O(\alpha_s)$ cross section into two pieces:

$$\frac{d\sigma}{dq_T^2\,dy}=X(q_T^2,Q^2,y)+Y(q_T^2,Q^2,y). \tag{10}$$

The terms in X contain integrable distributions which are singular at $q_T=0$. The function Y is perfectly finite as $q_T\to 0$. Up to and including $O(\alpha_s)$ the expression for X is ($C_F=\frac{4}{3}$)

$$\begin{aligned} X(q_T^2,Q^2,y)=N\Bigg\{&H(x_1^0,x_2^0)\big[\delta(q_T^2)(1+F(Q^2,y))+S(q_T^2,Q^2,y)\big]\\ &+\frac{\alpha_s}{2\pi}C_F\left(\frac{1}{(q_T^2)_+}+\delta(q_T^2)\left(\ln\left(\frac{A_T}{\mu}\right)^2-\frac{1}{\hat{\varepsilon}}\right)\right)\\ &\times\left[\int_{x_1^0}^1\frac{dz}{z}H(x_1^0/z,x_2^0)P_{qq}(z)+\int_{x_2^0}^1\frac{dz}{z}H(x_1^0,x_2^0/z)P_{qq}(z)\right]\\ &+\frac{\alpha_s}{2\pi}C_F\delta(q_T^2)\Bigg[\int_{x_1^0}^1\frac{dz}{z}c_q(z)H(x_1^0/z,x_2^0)\\ &+\int_{x_2^0}^1\frac{dz}{z}c_q(z)H(x_1^0,x_2^0/z)\Bigg]\Bigg\}. \end{aligned} \tag{11}$$

H is the product of quark and antiquark parton densities. For simplicity, we drop the terms due to initial gluons in this section. The singularity in $1/\hat{\varepsilon}$:

$$\frac{1}{\hat{\varepsilon}} = \left(\frac{2}{4-n} + \ln(4\pi) - \gamma_{\mathrm{E}}\right) \tag{12}$$

is a manifestation of the collinear divergence. The distribution S is given by

$$S(q_{\mathrm{T}}^2, Q^2, y) = \frac{\alpha_s}{2\pi} C_{\mathrm{F}} \left[2\left(\frac{\ln(Q^2/q_{\mathrm{T}}^2)}{q_{\mathrm{T}}^2}\right)_+ - \frac{3}{(q_{\mathrm{T}}^2)_+}\right]. \tag{13}$$

1.4. STEP IV

Perform the Fourier transform of X into impact parameter space and factor the collinear singularities into the parton distribution functions defined in deep inelastic scattering. The expression for X is now given by

$$X(b^2, Q^2, y) = N\Bigg\{H(x_1^0, x_2^0, P^2)\left[1 + F(Q^2, y) + S(b^2, Q^2, y)\right] + \frac{\alpha_s}{2\pi} C_{\mathrm{F}}\left[\int_{x_1^0}^1 \frac{\mathrm{d}z}{z} f_{\mathrm{q}}(z) H(x_1^0/z, x_2^0) + \int_{x_2^0}^1 \frac{\mathrm{d}z}{z} f_{\mathrm{q}}(z) H(x_1^0, x_2^0/z)\right]\Bigg\}, \tag{14}$$

where P is a b dependent scale,

$$S(b^2, Q^2, y) = \int_0^{A_{\mathrm{T}}^2} \frac{\mathrm{d}k^2}{k^2} \frac{\alpha_s(k^2)}{2\pi} C_{\mathrm{F}}(J_0(bk) - 1)\left(2\ln\frac{Q^2}{Qk^2} - 3\right), \tag{15}$$

and A_{T}^2 is the kinematic limit for the transverse momentum squared. The resummation is performed in b space by making the replacement

$$(1 + S(b^2, Q^2, y)) \to \exp S(b^2, Q^2, y). \tag{16}$$

Thus, X may be written as

$$X(b^2, Q^2, y) = R(b^2, Q^2, y) \exp S(b^2, Q^2, y). \tag{17}$$

1.5. STEP V

The differential q_{T} distribution is recovered by Fourier transforming back to q_{T} space:

$$\frac{\mathrm{d}\sigma}{\mathrm{d}q_{\mathrm{T}}^2 \mathrm{d}y} = \int \frac{\mathrm{d}^2\boldsymbol{b}}{4\pi} \mathrm{e}^{-i\boldsymbol{b}\cdot\boldsymbol{q}_{\mathrm{T}}}\left[R(b^2, Q^2, y) \exp S(b^2, Q^2, y)\right] + Y(q_{\mathrm{T}}^2, Q^2, y). \tag{18}$$

Note that upon integration over dq_T^2 the perturbative result for the total cross section $d\sigma/dy$ is obtained up to and including terms of order α_s. Similarly, the perturbative result for the average value of q_T^2 is also obtained.

2. Derivation of the q_T distribution

From ref. [5], we find that the graphs of figs. 1c and d give the following contribution to the quark–antiquark annihilation cross section in n dimensions ($n = 4 - 2\varepsilon$):

$$\frac{s}{\pi}\frac{d\sigma^{q\bar{q}}}{dt\,du} = N'\frac{\alpha_s}{2\pi}C_F\frac{1-\varepsilon}{\Gamma(1-\varepsilon)}\left(\frac{4\pi\mu^2}{q_T^2}\right)^{\varepsilon} M_1(s,t,u)\,\delta(s+t+u-Q^2), \quad (19)$$

where the matrix element squared is given by

$$M_1(s,t,u) = \frac{1}{s}\left\{(1-\varepsilon)\frac{(s-Q^2)^2}{ut} + \frac{2Q^2 s}{ut} - 2\right\}, \quad (20)$$

and N' is an overall normalization factor.

The corresponding contribution of the virtual diagrams figs. 1a and b is

$$\frac{s}{\pi}\frac{d\sigma^{q\bar{q}}}{dt\,du} = N'(1-\varepsilon)M_0(s)\left[1 + \frac{\alpha_s}{2\pi}C_F\frac{K}{\Gamma(1-\varepsilon)}\left(\frac{4\pi\mu^2}{Q^2}\right)^{\varepsilon}\right]\delta(s+t+u-Q^2),$$

(21)

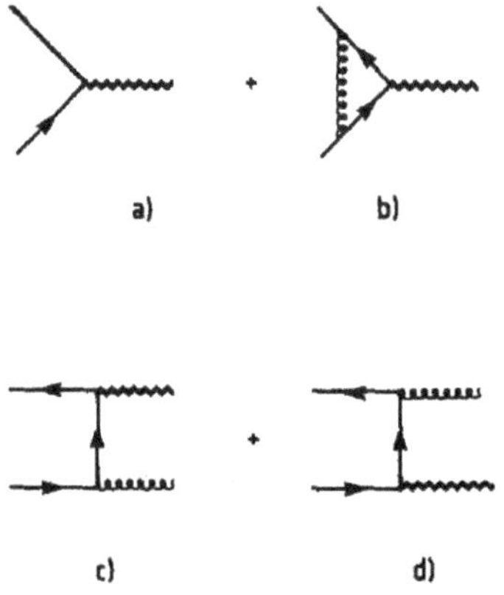

Fig. 1. Feynman diagrams for the $q\bar{q}$ annihilation process up to and including terms of order α_s. The produced vector boson is denoted by a wavy line and the gluon by a curly line.

where

$$M_0(s)=\delta(s-Q^2), \qquad K=\left(-\frac{2}{\varepsilon^2}-\frac{3}{\varepsilon}+\pi^2-8\right). \tag{22}$$

The substitution of these parton cross sections into eq. (4) involves integrals of the general form:

$$I=\int dx_1 dx_2 f(x_1,x_2)\delta\big(x_1x_2S+x_1(T-Q^2)+x_2(U-Q^2)+Q^2\big), \tag{23}$$

where the delta function present in the parton cross sections equations (19) and (21) has been expressed in hadronic variables. Performing the x_2 integration and splitting the x_1 range of integration, we find

$$I=\int_{\sqrt{\tau_+}e^y}^1 \frac{dx_1 f(x_1,x_2^*)}{x_1S+U-Q^2}+\int_{\sqrt{\tau_+}e^{-y}}^1 \frac{dx_2 f(x_1^*,x_2)}{x_2S+T-Q^2}, \tag{24}$$

where x_1^* and x_2^* are fixed to be

$$x_1^*=\frac{x_2(Q^2-U)-Q^2}{x_2S+T-Q^2}, \qquad x_2^*=\frac{x_1(Q^2-T)-Q^2}{x_1S+U-Q^2}. \tag{25}$$

The variable τ_+ is

$$\sqrt{\tau_+}=\sqrt{\frac{q_T^2}{S}}+\sqrt{\tau+\frac{q_T^2}{S}}. \tag{26}$$

In performing the x_1 and x_2 integrations in eq. (24), a certain proliferation of variables is inevitable. For the convenience of the reader, we have collected their definitions in table 1. To lighten the notation we shall drop the superscript on x_1^*, x_2^* in the following. We rewrite eq. (24) as

$$I=\frac{1}{S}\left\{\int_{\sqrt{\tau_+}e^y}^1 \frac{dx_1 f(x_1,x_2)}{x_1-x_1^+}+\int_{\sqrt{\tau_+}e^{-y}}^1 \frac{dx_2 f(x_1,x_2)}{x_2-x_2^+}\right\}. \tag{27}$$

Inserting the parton cross section equation (19) into eq. (23), the hadronic cross section in n dimensions is given by

$$\frac{d\sigma}{dq_T^2\,dy}=N\frac{\alpha_s}{2\pi}C_F\frac{1-\varepsilon}{\Gamma(1-\varepsilon)}\left(\frac{4\pi\mu^2}{q_T^2}\right)^{\varepsilon}$$

$$\times\int dx_1 dx_2 M_1(s,t,u)\delta\big(x_1x_2-x_1x_2^+-x_2x_1^++x_1^0x_2^0\big), \tag{28}$$

TABLE 1

$p(P_1)+\bar{p}(P_2)\to B(q)+X$

$S=(P_1+P_2)^2;\quad T=(P_1-q)^2;\quad U=(P_2-q)^2;\quad q^2=Q^2;\quad \tau=Q^2/S$

$y=\frac{1}{2}\ln\frac{q^0+q^3}{q^0-q^3};\quad (T-Q^2)=-\sqrt{S}\left(Q^2+q_T^2\right)^{1/2}e^{-y};\quad (U-Q^2)=-\sqrt{S}\left(Q^2+q_T^2\right)^{1/2}e^{y}$

$x_1^0=\sqrt{\tau}\,e^{y};\quad x_2^0=\sqrt{\tau}\,e^{-y};\quad x_1^+=\frac{(Q^2-U)}{S};\quad x_2^+=\frac{(Q^2-T)}{S},\quad x_1^*=\frac{x_2x_1^+-\tau}{x_2-x_2^+};$

$x_2^*=\frac{x_1x_2^+-\tau}{x_1-x_1^+};$

$A_T^2=\left[\frac{(S+Q^2)^2}{4S\cosh^2 y}-Q^2\right]=Q^2\frac{\left(1-x_1^{0^2}\right)\left(1-x_2^{0^2}\right)}{\left(x_1^0+x_2^0\right)^2},\quad \sqrt{\tau_\pm}=\sqrt{q_T^2/S}\pm\left(\tau+q_T^2/S\right)^{1/2}$

$s=(p_1+p_2)^2=x_1x_2S;\quad t=(p_1-q)^2=x_1(T-Q^2)+Q^2;\quad u=(p_2-q)^2=x_2(U-Q^2)+Q^2$

and the matrix element squared expressed in hadronic variables is

$$M_1=\left\{(1-\varepsilon)\frac{(1-\tau/x_1x_2)^2}{q_T^2}+\frac{2\tau}{x_1x_2q_T^2}-\frac{2}{x_1x_2S}\right\}. \tag{29}$$

By addition and subtraction of the residues at $q_T^2=0$, we can isolate the terms which are singular in the limit $q_T^2\to 0$. We treat the x_1 and x_2 integrals resulting from the application of eq. (27) in eq. (28) separately. The x_1 integral may be written (dropping overall factors) as

$$\begin{aligned}
&\int_{\sqrt{\tau_+}e^y}^1\frac{dx_1}{x_1-x_1^+}H(x_1,x_2)\left\{(1-\varepsilon)\left(1-\frac{\tau}{x_1x_2}\right)^2\frac{1}{q_T^2}+\frac{2\tau}{x_1x_2q_T^2}-\frac{2}{x_1x_2S}\right\}\left(\frac{1}{q_T^2}\right)^{\varepsilon}\\
&\equiv\int_{\sqrt{\tau_+}e^y}^1\frac{dx_1}{x_1-x_1^+}\left(\frac{1}{q_T^2}\right)^{1+\varepsilon}\left\{H(x_1,x_2)\left[(1-\varepsilon)\left(1-\frac{\tau}{x_1x_2}\right)^2+\frac{2\tau}{x_1x_2}\right]\right.\\
&\qquad\left.-2H(x_1^0,x_2^0)\right\}\\
&\quad+2H(x_1^0,x_2^0)\left(\frac{1}{q_T^2}\right)^{1+\varepsilon}\int_{\sqrt{\tau_+}e^y}^1\frac{dx_1}{x_1-x_1^+}-2\int_{\sqrt{\tau_+}e^y}^1\frac{dx_1}{x_1-x_1^+}\frac{H(x_1,x_2)}{x_1x_2S}.
\end{aligned} \tag{30}$$

The step performed in the above equation must be made before subtracting the residue of the pole at $q_T^2=0$. Otherwise, one would be led to an unregulated divergence at $x_{1,2}=x_{1,2}^0$, because at $q_T=0$, $x_{1,2}^+=x_{1,2}^0$, $\sqrt{\tau_+}\,e^{\pm y}=x_{1,2}^0$.

A similar formula can be derived for the x_2 part of the integral. We now isolate the poles in q_T^2 by further subtraction and use of the identity

$$\int_0^{A_T^2} \mathrm{d}q_T^2 \frac{f(q_T^2)}{(q_T^2)^{1+\varepsilon}} \equiv \int_0^{A_T^2} \mathrm{d}q_T^2 \frac{f(q_T^2)-f(0)}{(q_T^2)^{1+\varepsilon}} + f(0)\int_0^{A_T^2} \frac{\mathrm{d}q_T^2}{(q_T^2)^{1+\varepsilon}}, \tag{31}$$

where A_T^2 is the kinematic limit of the transverse momentum squared (cf. table 1). Eq. (31) can formally be written as

$$\frac{1}{(q_T^2)^{1+\varepsilon}} \equiv \frac{1}{(q_T^2)_+} + \left(\ln A_T^2 - \frac{1}{\varepsilon}\right)\delta(q_T^2) + \mathrm{O}(\varepsilon). \tag{32}$$

Applying this separation to the first term in eq. (30) we obtain

$$\begin{aligned} &\frac{1}{(q_T^2)_+} \int_{\sqrt{\tau_+}e^y}^1 \frac{\mathrm{d}x_1}{x_1 - x_1^+} \left\{ H(x_1, x_2)\left[\left(1 - \frac{\tau}{x_1 x_2}\right)^2 + \frac{2\tau}{x_1 x_2}\right] - 2H(x_1^0, x_2^0) \right\} \\ &+ \delta(q_T^2)\left(\ln A_T^2 - \frac{1}{\varepsilon}\right)\int_{x_1^0}^1 \frac{\mathrm{d}x_1}{x_1 - x_1^0}\left\{ H(x_1, x_2^0)\left[(1-\varepsilon)\left(1 - \frac{x_1^0}{x_1}\right)^2 + \frac{2x_1^0}{x_1}\right] \right. \\ &\left. - 2H(x_1^0, x_2^0) \right\}. \end{aligned} \tag{33}$$

When applying the same procedure to the second term in eq. (30), it is convenient to consider the x_1 and x_2 integrals simultaneously:

$$\begin{aligned} &2H(x_1^0, x_2^0)\frac{1}{(q_T^2)^{1+\varepsilon}}\left[\int_{\sqrt{\tau_+}e^y}^1 \frac{\mathrm{d}x_1}{x_1 - x_1^+} + \int_{\sqrt{\tau_+}e^{-y}}^1 \frac{\mathrm{d}x_2}{x_2 - x_2^+}\right] \\ &= 2H(x_1^0, x_2^0)\frac{1}{(q_T^2)^{1+\varepsilon}} \ln\frac{(1-x_1^+)(1-x_2^+)S}{q_T^2}. \end{aligned} \tag{34}$$

By use of eq. (32) and the identity

$$\frac{1}{(q_T^2)^{1+\varepsilon}}\ln q_T^2 = \left(\frac{\ln q_T^2}{q_T^2}\right)_+ + \left(\tfrac{1}{2}\ln^2 A_T^2 - \frac{1}{\varepsilon^2}\right)\delta(q_T^2) + \mathrm{O}(\varepsilon), \tag{35}$$

eq. (34) may be written as

$$\begin{aligned} &2H(x_1^0, x_2^0)\left\{\frac{1}{(q_T^2)_+}\ln[(1-x_1^+)(1-x_2^+)S] - \left(\frac{\ln q_T^2}{q_T^2}\right)_+\right\} \\ &+ 2H(x_1^0, x_2^0)\delta(q_T^2)\left\{\left(\ln A_T^2 - \frac{1}{\varepsilon}\right)\ln[(1-x_1^0)(1-x_2^0)S] + \frac{1}{\varepsilon^2} - \tfrac{1}{2}\ln^2 A_T^2\right\}. \end{aligned} \tag{36}$$

The third term in eq. (30) is completely finite as $q_T \to 0$ and requires no subtraction. Three further manipulations complete step II:

(i) Addition of the contributions from the x_2 part of the integral where not already included.

(ii) Addition of the virtual contribution of eq. (21). After dropping the lowest-order term and the same overall factor as in the above, the virtual contribution is

$$H(x_1^0, x_2^0)\delta(q_T^2)\left(-\frac{2}{\varepsilon^2}+\frac{2}{\varepsilon}\ln\frac{Q^2}{\mu^2}-\ln^2\frac{Q^2}{\mu^2}-\frac{3}{\varepsilon}+3\ln\frac{Q^2}{\mu^2}+\pi^2-8\right). \quad (37)$$

(iii) Simplification of the formula by use of the identity

$$\int_{x_1^0}^{1}\frac{dz}{z}\frac{f(z)-f(1)}{1-z}=\int_{x_1^0}^{1}\frac{dz}{z}f(z)\frac{1}{(1-z)_+}+f(1)\ln\frac{x_1^0}{1-x_1^0}. \quad (38)$$

The "+" distribution in the above formula is defined on the range 0 to 1 such that

$$\int_0^1 dz\,\frac{f(z)}{(1-z)_+}=\int_0^1 dz\,\frac{f(z)-f(1)}{1-z}. \quad (39)$$

Upon completion of these manipulations we obtain

$$\begin{aligned}\frac{d\sigma}{dq_T^2\,dy}=N\frac{\alpha_s}{2\pi}C_F\Bigg\{&H(x_1^0,x_2^0)\left[\frac{2}{(q_T^2)_+}\ln((1-x_1^+)(1-x_2^+)S)-2\left(\frac{\ln q_T^2}{q_T^2}\right)_+\right.\\&\left.+\left(3\ln\frac{Q^2}{A_T^2}-\ln^2\frac{Q^2}{A_T^2}\right)\delta(q_T^2)\right]\\&+\frac{1}{(q_T^2)_+}\left[\int_{\sqrt{\tau_+}e^{y}}^{1}\frac{dx_1}{x_1-x_1^+}+\int_{\sqrt{\tau_+}e^{-y}}^{1}\frac{dx_2}{x_2-x_2^+}\right]\\&\times\left[H(x_1,x_2)\left(1+\left(\frac{\tau}{x_1x_2}\right)^2\right)-2H(x_1^0,x_2^0)\right]\\&+\delta(q_T^2)\left(\ln\left(\frac{A_T}{\mu}\right)^2-\frac{1}{\hat{\varepsilon}}\right)\left[\int_{x_1^0}^{1}\frac{dz}{z}P_{qq}(z)H(x_1^0/z,x_2^0)\right.\\&\left.+\int_{x_2^0}^{1}\frac{dz}{z}P_{qq}(z)H(x_1^0,x_2^0/z)\right]\\&+\delta(q_T^2)\left[\int_{x_1^0}^{1}\frac{dz}{z}c_q(z)H(x_1^0/z,x_2^0)+\int_{x_2^0}^{1}\frac{dz}{z}c_q(z)H(x_1^0,x_2^0/z)\right]\\&-\left[\int_{\sqrt{\tau_+}e^{y}}^{1}\frac{dx_1}{x_1-x_1^+}+\int_{\sqrt{\tau_+}e^{-y}}^{1}\frac{dx_2}{x_2-x_2^+}\right]\frac{2H(x_1,x_2)}{x_1x_2S}\Bigg\}, \end{aligned} \quad (40)$$

where

$$P_{qq}(z)=\frac{1+z^2}{(1-z)_+}+\frac{3}{2}\delta(1-z),$$

$$c_q(z)=(1-z)+\delta(1-z)\left(\tfrac{1}{2}\pi^2-4\right),$$

$$\frac{1}{\hat{\varepsilon}}=\frac{1}{\varepsilon}+\ln(4\pi)-\gamma_E. \tag{41}$$

We can finally write the result of the order α_s $q\bar{q}$ annihilation term:

$$\frac{d\sigma}{dq_T^2\,dy}=X_q(q_T^2,Q^2,y)+Y_q(q_T^2,Q^2,y), \tag{42}$$

where

$$X_q(q_T^2,Q^2,y)=X_q^{(0)}+X_q^{(1)},$$

$$X_q^{(0)}=NH(x_1^0,x_2^0)\delta(q_T^2),$$

$$\begin{aligned}X_q^{(1)}=N\frac{\alpha_s}{2\pi}C_F\Bigg\{&H(x_1^0,x_2^0)\left[2\left(\frac{\ln(Q^2/q_T^2)}{q_T^2}\right)_+-\frac{3}{(q_T^2)_+}\right.\\&\left.+\left(-3\ln\frac{A_T^2}{Q^2}-\ln^2\frac{A_T^2}{Q^2}\right)\delta(q_T^2)\right]\\&+\left[\frac{1}{(q_T^2)_+}+\delta(q_T^2)\left(\ln\frac{A_T^2}{\mu^2}-\frac{1}{\hat{\varepsilon}}\right)\right]\\&\times\left[\int_{x_1^0}^1\frac{dz}{z}H(x_1^0/z,x_2^0)P_{qq}(z)+\int_{x_2^0}^1\frac{dz}{z}H(x_1^0,x_2^0/z)P_{qq}(z)\right]\\&+\delta(q_T^2)\left[\int_{x_1^0}^1\frac{dz}{z}c_q(z)H(x_1^0/z,x_2^0)\right.\\&\left.\left.+\int_{x_2^0}^1\frac{dz}{z}c_q(z)H(x_1^0,x_2^0/z)\right]\right\}.\end{aligned} \tag{43}$$

The + distributions in the above equation are defined with respect to the upper limit of the q_T^2 integration A_T^2. A different choice B^2 for the upper limit would modify the distributions such that

$$\frac{1}{(q_T^2)_+} \to \frac{1}{(q_T^2)_{B_+}} + \ln\frac{B^2}{A_T^2}\delta(q_T^2),$$

$$\left(\frac{\ln(Q^2/q_T^2)}{q_T^2}\right)_+ \to \left(\frac{\ln(Q^2/q_T^2)}{q_T^2}\right)_{B_+} + \tfrac{1}{2}\ln\frac{B^2}{A_T^2}\ln\left(\frac{Q^4}{A_T^2B^2}\right)\delta(q_T^2). \tag{44}$$

The second term in eq. (42), Y_q contains only terms which are finite in the limit $q_T^2 \to 0$. The explicit form of Y_q will be given in the next section. This completes step III.

The fourth step is to perform the Fourier transform of X_q to impact parameter space. The expression for X_q becomes

$$\begin{aligned} X_q(b^2,Q^2,y) = N\Big\{ & H(x_1^0,x_2^0)\left[1+F(Q^2,y)+S(b^2,Q^2,y)\right] \\ & + \frac{\alpha_s}{2\pi}C_F\left(\ln\frac{P^2}{\mu}-\frac{1}{\hat{\varepsilon}}\right)\left[\int_{x_1^0}^1\frac{dz}{z}H(x_1^0/z,x_2^0)P_{qq}(z)\right. \\ & \left. + \int_{x_2^0}^1\frac{dz}{z}H(x_1^0,x_2^0/z)P_{qq}(z)\right] \\ & + \frac{\alpha_s}{2\pi}C_F\left[\int_{x_1^0}^1\frac{dz}{z}c_q(z)H(x_1^0/z,x_2^0)\right. \\ & \left. + \int_{x_2^0}^1\frac{dz}{z}c_q(z)H(x_1^0,x_2^0/z)\right]\Big\}, \end{aligned} \tag{45}$$

where

$$F(Q^2,y) = \frac{\alpha_s}{2\pi}C_F\left[-3\ln\frac{A_T^2}{Q^2} - \ln^2\frac{A_T^2}{Q^2}\right], \tag{46}$$

$$\ln P^2 = \ln A_T^2 + \int_0^{A_T^2}\frac{d^2q_T}{\pi}\left(e^{ib\cdot q_T}-1\right)\frac{1}{q_T^2}, \tag{47}$$

and $S(b)$ is given by eq. (15). The collinear singularities are factored and the finite terms are fixed in terms of the deep inelastic structure function F_2 at the scale Q^2:

$$q(x,Q^2)=q_0(x)+\int_x^1\frac{dz}{z}\frac{\alpha_s}{2\pi}C_F\left[\left(\ln\frac{Q^2}{\mu^2}-\frac{1}{\hat{\varepsilon}}\right)P_{qq}(z)+c_q^{(2)}(z)\right]q_0\left(\frac{x}{z}\right)$$

$$+\int_x^1\frac{dz}{z}\frac{\alpha_s}{2\pi}T_R\left[\left(\ln\frac{Q^2}{\mu^2}-\frac{1}{\hat{\varepsilon}}\right)P_{qg}(z)+c_g^{(2)}(z)\right]g_0\left(\frac{x}{z}\right), \tag{48}$$

where $T_R=\frac{1}{2}$ and [5]

$$c_q^{(2)}(z)=\left\{(1+z^2)\left(\frac{\ln(1-z)}{1-z}\right)_+-\frac{3}{2}\frac{1}{(1-z)_+}-\frac{1+z^2}{1-z}\ln z+3\right.$$

$$\left.+2z-\left(\tfrac{9}{2}+\tfrac{1}{3}\pi^2\right)\delta(1-z)\right\},$$

$$c_g^{(2)}(z)=\left\{\left[z^2+(1-z)^2\right]\ln\frac{1-z}{z}+6z(1-z)\right\}. \tag{49}$$

After resummation of the form factor S we obtain in b space:

$$X_q(b^2,Q^2,y)=N\left\{\exp S(b^2,Q^2,y)\left[H(x_1^0,x_2^0,P^2)(1+F(Q^2,y))\right]\right.$$

$$\left.+\frac{\alpha_s}{2\pi}C_F\left[\int_{x_1^0}^1\frac{dz}{z}f_q(z)H(x_1^0/z,x_2^0)+\int_{x_2^0}^1\frac{dz}{z}f_q(z)H(x_1^0,x_2^0/z)\right]\right\}, \tag{50}$$

where

$$f_q(z)=c_q(z)-c_q^{(2)}(z). \tag{51}$$

The gluon term in eq. (48) will be included in the Compton contribution which we now consider.

The result from the quark–gluon scattering (fig. 2) is given by

$$\frac{s}{\pi}\frac{d\sigma^{qg}}{dt\,du}=N'\frac{\alpha_s}{2\pi}T_R\frac{1-\varepsilon}{\Gamma(1-\varepsilon)}\left(\frac{4\pi\mu^2}{q_T^2}\right)^{\varepsilon}M_2(s,t,u)\delta(s+t+u-Q^2), \tag{52}$$

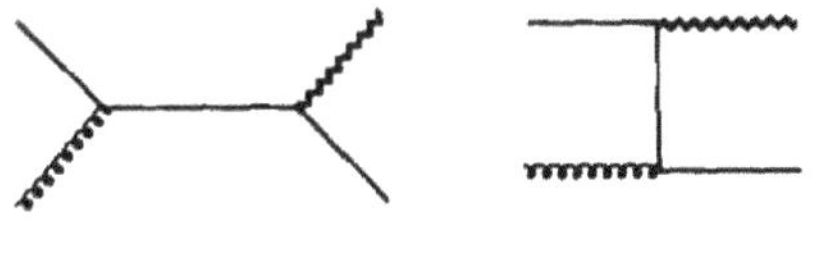

Fig. 2. The production of a vector boson from an initial state gluon.

where the matrix element squared is

$$M_2(s,t,u)=\left\{\frac{1}{q_T^2}\left[-(1-\varepsilon)\frac{u}{s}-\frac{2u^2Q^2}{s^3}\right]-(1-\varepsilon)\frac{t}{s^2}\right\}. \tag{53}$$

Proceeding through the first three steps we easily obtain

$$\begin{aligned}\frac{\mathrm{d}\sigma}{\mathrm{d}q_T^2\,\mathrm{d}y}=N\frac{\alpha_s}{2\pi}T_R\Bigg\{\Bigg[\Bigg[\int \mathrm{d}x_1\,\mathrm{d}x_2\,\delta\left(x_1x_2-x_1x_2^+-x_2x_1^++x_1^0x_2^0\right)K_1(x_1,x_2)\\ \times\left[\frac{1}{q_T^2}\left(\frac{x_2x_1^+-x_1^0x_2^0}{x_1x_2}-\frac{2x_1^0x_2^0\left(x_2x_1^+-x_1^0x_2^0\right)^2}{(x_1x_2)^3}\right)\right.\\ \left.+\frac{x_1x_2^+-x_1^0x_2^0}{(x_1x_2)^2S}\right]\\ -\int_{x_2^0}^1\mathrm{d}x_2\,K_1(x_1^0,x_2)\frac{1}{q_T^2}\left(\frac{1}{x_2}-\frac{2(x_2-x_2^0)x_2^0}{x_2^3}\right)\Bigg]+[(1\leftrightarrow 2)]\\ +\left[\frac{1}{(q_T^2)_+}+\ln\frac{A_T^2}{\mu^2}-\frac{1}{\hat\varepsilon}\right]\left[\int_{x_2^0}^1\frac{\mathrm{d}z}{z}P_{qg}(z)K_1(x_1^0,x_2^0/z)\right.\\ \left.+\int_{x_1^0}^1\frac{\mathrm{d}z}{z}P_{qg}(z)K_2(x_1^0/z,x_2^0)\right]\\ +\delta(q_T^2)\left[\int_{x_2^0}^1\frac{\mathrm{d}z}{z}K_1(x_1^0,x_2^0/z)c_g(z)+\int_{x_1^0}^1\frac{\mathrm{d}z}{z}c_g(z)K_2(x_1^0/z,x_2^0)\right]\Bigg\},\end{aligned} \tag{54}$$

where

$$\begin{aligned}P_{qg}(z)&=\left[z^2+(1-z)^2\right],\qquad c_g(z)=1,\\ K_1(x_1,x_2)&\sim\left[q(x_1)+\bar q(x_1)\right]g(x_2),\\ K_2(x_1,x_2)&\sim\left[q(x_2)+\bar q(x_2)\right]g(x_1).\end{aligned} \tag{55}$$

Eq. (54) is written for proton (anti-)proton scattering. With obvious modifications to take into account the differing distributions of quarks in pions and protons, it would also be valid for πP scattering.

3. Analytic results

The final analytic results suitable for use in numerical calculations are given here. The results form a self-contained set which can be used almost without reference to previous sections. The principal analytic result of our paper is the expression for the cross section for the production of a vector boson:

$$\frac{\mathrm{d}\sigma}{\mathrm{d}q_{\mathrm{T}}^2\,\mathrm{d}y} = N\left\{\int \frac{\mathrm{d}^2 \boldsymbol{b}}{4\pi} \mathrm{e}^{-i\boldsymbol{q}_{\mathrm{T}}\cdot \boldsymbol{b}} \exp S(b^2, Q^2, y) R(b^2, Q^2, y) + Y(q_{\mathrm{T}}^2, Q^2, y)\right\}, \tag{56}$$

where

$$\begin{aligned} R(b^2, Q^2, y) = {} & H(x_1^0, x_2^0, P^2)\left[1 + \frac{\alpha_s}{2\pi}\frac{4}{3}\left(-3\ln\frac{A_{\mathrm{T}}^2}{Q^2} - \ln^2\frac{A_{\mathrm{T}}^2}{Q^2}\right)\right] \\ & + \frac{\alpha_s}{2\pi}\frac{4}{3}\left[\int_{x_1^0}^1 \frac{\mathrm{d}z}{z} f_{\mathrm{q}}(z) H(x_1^0/z, x_2^0, P^2)\right. \\ & \left. + \int_{x_2^0}^1 \frac{\mathrm{d}z}{z} f_{\mathrm{q}}(z) H(x_1^0, x_2^0/z, P^2)\right] \\ & + \frac{\alpha_s}{2\pi}\frac{1}{2}\left[\int_{x_1^0}^1 \frac{\mathrm{d}z}{z} f_{\mathrm{g}}(z) K_2(x_1^0/z, x_2^0, P^2)\right. \\ & \left. + \int_{x_2^0}^1 \frac{\mathrm{d}z}{z} f_{\mathrm{g}}(z) K_1(x_1^0, x_2^0/z, P^2)\right]. \end{aligned} \tag{57}$$

In this formula the scale at which the parton densities in H, K_1 and K_2 are probed is given by

$$\ln P^2 = \left[\ln A_T^2 + \int_0^{A_T^2} \frac{d^2 q_T}{\pi} (e^{-ib\cdot q_T} - 1)\frac{1}{q_T^2}\right] - \ln\frac{b_0^2}{b^2}, \tag{58}$$

where $b_0^2 = 4e^{-2\gamma_E} \sim 1.261$ [19] and the correction terms are defined as

$$f_q(z) = \left\{\frac{3}{2}\frac{1}{(1-z)_+} - (1+z^2)\left(\frac{\ln(1-z)}{1-z}\right)_+ + \frac{1+z^2}{1-z}\ln z - 2 - 3z + \left(\tfrac{1}{2} + \tfrac{5}{6}\pi^2\right)\delta(1-z)\right\},$$

$$f_g(z) = \left\{1 - 6z(1-z) - \left[z^2 + (1-z)^2\right]\ln\frac{1-z}{z}\right\}. \tag{59}$$

The quark distributions used in the above formula are defined beyond the leading order in terms of the deep inelastic structure function F_2 at the scale Q^2 and are given in eqs. (66). The form factor S is obtained by Fourier transforming the expression in eq. (13):

$$S(b^2, Q^2, y) = \int_0^{A_T^2} dq^2 \,[J_0(bq) - 1]\frac{\alpha_s(q^2)}{2\pi}\frac{4}{3}\left[2\frac{\ln(Q^2/q^2)}{q^2} - \frac{3}{q^2}\right]. \tag{60}$$

Note that there is no ambiguity in the value of the upper limit of integration A_T^2. Given the total cross section constraint a change of A_T^2 should be compensated by a corresponding change of R in eq. (57). For example, if one prefers Q^2 as an upper limit in the exponential, the replacement

$$\exp\int_0^{A_T^2} = \left(1 + \int_{Q^2}^{A_T^2}\right)\exp\int_0^{Q^2} \tag{61}$$

can be made. This is allowed because $\alpha_s(q_T^2)$ is small for $Q^2 < q_T^2 < A_T^2$. The resulting additional contribution to R cancels in this case the logarithms appearing in eq. (57) in the large-b limit. Actually, the replacement in eq. (61) is demanded when $A_T^2 \gg Q^2$, as for example is the case of W production at tevatron energies, in order to keep under numerical control the large $\log(A_T^2/Q^2)$ terms.

The residual finite term Y can be divided into the parts due to the annihilation and Compton scattering graphs:

$$Y(q_T^2, Q^2, y) = \frac{\alpha_s}{2\pi}\frac{4}{3}Y_q(q_T^2, Q^2, y) + \frac{\alpha_s}{2\pi}\frac{1}{2}Y_g(q_T^2, Q^2, y), \tag{62}$$

where

$$Y_{\rm q}(q_{\rm T}^2, Q^2, y) = -\frac{2}{S}\left[\int_{\sqrt{\tau_+}\,{\rm e}^{y}}^{1}\frac{{\rm d}x_1}{(x_1 - x_1^+)}\frac{H(x_1, x_2^*)}{x_1 x_2^*} + \int_{\sqrt{\tau_+}\,{\rm e}^{-y}}^{1}\frac{{\rm d}x_2}{(x_2 - x_2^+)}\frac{H(x_1^*, x_2)}{x_1^* x_2}\right]$$

$$+\frac{1}{q_{\rm T}^2}\left\{\int_{\sqrt{\tau_+}\,{\rm e}^{y}}^{1}\frac{{\rm d}x_1}{(x_1 - x_1^+)}\left[H(x_1, x_2^*)\left(1 + \left(\frac{\tau}{x_1 x_2^*}\right)^2\right) - 2H(x_1^0, x_2^0)\right]\right.$$

$$-\int_{x_1^0}^{1}\frac{{\rm d}x_1}{(x_1 - x_1^0)}\left[H(x_1, x_2^0)\left(1 + \left(\frac{x_1^0}{x_1}\right)^2\right) - 2H(x_1^0, x_2^0)\right]$$

$$+\int_{\sqrt{\tau_+}\,{\rm e}^{-y}}^{1}\frac{{\rm d}x_2}{(x_2 - x_2^+)}\left[H(x_1^*, x_2)\left(1 + \left(\frac{\tau}{x_1^* x_2}\right)^2\right) - 2H(x_1^0, x_2^0)\right]$$

$$-\int_{x_1^0}^{1}\frac{{\rm d}x_2}{(x_2 - x_2^0)}\left[H(x_1^0, x_2)\left(1 + \left(\frac{x_2^0}{x_2}\right)^2\right) - 2H(x_1^2, x_2^0)\right]$$

$$\left. + 2H(x_1^0, x_2^0)\ln\frac{(1 - x_1^+)(1 - x_2^+)}{(1 - x_1^0)(1 - x_2^0)}\right\}, \tag{63}$$

$$Y_{\rm g}(q_{\rm T}^2, Q^2, y) = \left\{\frac{1}{q_{\rm T}^2}\int_{\sqrt{\tau_+}\,{\rm e}^{y}}^{1}\frac{{\rm d}x_1}{(x_1 - x_1^+)}K_1(x_1, x_2^*)\left[\frac{x_2^* x_1^+ - \tau}{x_1 x_2^*} - \frac{2\tau(x_2^* x_1^+ - \tau)^2}{(x_1 x_2^*)^3}\right]\right.$$

$$+\frac{1}{q_{\rm T}^2}\int_{\sqrt{\tau_+}\,{\rm e}^{-y}}^{1}\frac{{\rm d}x_2}{(x_2 - x_2^+)}K_1(x_1^*, x_2)\left[\frac{x_2 x_1^+ - \tau}{x_1^* x_2} - \frac{2\tau(x_2 x_1^+ - \tau)^2}{(x_1^* x_2)^3}\right]$$

$$-\frac{1}{q_{\rm T}^2}\int_{x_2^0}^{1}\frac{{\rm d}x_2}{x_2}K_1(x_1^0, x_2)\left[1 - 2\frac{x_2^0}{x_2}\left(1 - \frac{x_2^0}{x_2}\right)\right]$$

$$+\frac{1}{S}\left[\int_{\sqrt{\tau_+}\,{\rm e}^{y}}^{1}\frac{{\rm d}x_1}{(x_1 - x_1^+)}K_1(x_1, x_2^*)\frac{x_1 x_2^+ - \tau}{(x_1 x_2^*)^2}\right.$$

$$\left.\left. + \int_{\sqrt{\tau_+}\,{\rm e}^{-y}}^{1}\frac{{\rm d}x_2}{(x_2 - x_2^+)}K_1(x_1^*, x_2)\frac{x_1^* x_2^+ - \tau}{(x_1^* x_2)^2}\right] + (1 \leftrightarrow 2)\right\}, \tag{64}$$

where $x_{1,2}$, $x_{1,2}^+$, $x_{1,2}^*$, $\sqrt{\tau_+}$ are defined in table 1 and the exchange $(1 \leftrightarrow 2)$ also includes $K_1(x_1, x_2^*) \leftrightarrow K_2(x_1^*, x_2)$, $K_1 x_1^*, x_2) \leftrightarrow K_2(x_1, x_2^*)$, $K_1(x_1^0, x_2) \leftrightarrow K_2(x_1, x_2^0)$ and $y \leftrightarrow -y$.

The overall normalization factors in eq. (56) for the production of the three types of vector bosons are given by

$$N(\gamma^*) = \frac{4\pi^2\alpha}{3S}, \qquad N(Z) = \frac{\pi^2\alpha_w}{12S\cos^2\theta_w}, \qquad N(W) = \frac{\pi^2\alpha_w}{3S}, \qquad (65)$$

and $\alpha_w = \alpha/\sin^2\theta_w$. The differential cross section $d\sigma/dQ^2 dq_T^2 dy$ usually quoted for the production of a lepton pair of invariant mass Q is obtained by multiplying $N(\gamma^*)$ by a further factor of $(\alpha/3\pi Q^2)$. We also define the products of the parton distribution functions:

$$H^\gamma(x_1, x_2, Q^2) = \sum_f e_f^2 \{ q_f(x_1, Q^2)\bar{q}_f(x_2, Q^2) + (1 \leftrightarrow 2) \},$$

$$H^Z(x_1, x_2, Q^2) = \sum_f n_f^z \{ q_f(x_1, Q^2)\bar{q}_f(x_2, Q^2) + (1 \leftrightarrow 2) \},$$

$$H^{W^-}(x_1, x_2, Q^2) = \{ [u(x_1, Q^2)\bar{d}(x_2, Q^2) + c(x_1, Q^2)\bar{s}(x_2, Q^2)]\cos^2\theta_c$$

$$+ [u(x_1, Q^2)\bar{s}(x_2, Q^2) + c(x_1, Q^2)\bar{d}(x_2, Q^2)]\sin^2\theta_c \},$$

$$(66)$$

and a similar expression for H^{W^-}. A quark of flavour f has a charge e_f, and

$$n_f^z = \left[(1 - 4|e_f|\sin^2\theta_w)^2 + 1 \right]. \qquad (67)$$

The analogous results involving the gluon distribution functions are

$$K_1^\gamma(x_1, x_2, Q^2) = \sum_f e_f^2 [q_f(x_1, Q^2) + \bar{q}_f(x_1, Q^2)] g(x_2, Q^2),$$

$$K_1^Z(x_1, x_2, Q^2) = \sum_f n_f^z [q_f(x_1, Q^2) + \bar{q}_f(x_1, Q^2)] g(x_2, Q^2),$$

$$K_1^{W^+}(x_1, x_2, Q^2) = [u(x_1, Q^2) + c(x_1, Q^2) + \bar{d}(x_1, Q^2) + \bar{s}(x_1, Q^2)] g(x_2, Q^2),$$

$$(68)$$

while $K_2^i(x_1, x_2, Q^2)$ are obtained by interchanging quark and gluon densities

$(K_2 \sim q(x_2)g(x_1))$. In eqs. (62)–(64), no scale is written for α_s and the product of the parton densities in $Y_{q,g}$ which is undetermined to this order in α_s. We have taken both these scales to be Q^2. A different choice leads to numerical variations of the results which have to be taken into account in the estimate of the theoretical error.

Our result for the form factor is in agreement with the general formalism of Collins and Soper [12,13] in the leading DLA. After some manipulations their result for the form factor can be written as

$$S_{\mathrm{CS}}(b^2, Q^2) = -\frac{C_F}{\pi}\int_{c_1^2/b^2}^{c_2^2 Q^2} \frac{\mathrm{d}k^2}{k^2}\left[\ln\frac{c_2^2 Q^2}{k^2} A(k^2) + B(k^2)\right], \tag{69}$$

where in the $\overline{\mathrm{MS}}$ scheme:

$$A(k^2) = \left[\alpha_s(k^2) + D\alpha_s^2(k^2) - \ln\frac{c_1^2}{b_0^2}\frac{\mathrm{d}}{\mathrm{d}\ln k^2}\alpha_s(k^2)\right] + \mathrm{O}(\alpha_s^3),$$

$$B(k^2) = \alpha_s(k^2)\left[-\tfrac{3}{2} + \ln\frac{c_1^2}{b_0^2} - \ln c_2^2\right] + \mathrm{O}(\alpha_s^2), \tag{70}$$

and c_1 and c_2 are arbitrary constants. A change in c_1 or c_2 in the form factor is exactly compensated by corresponding variations elsewhere in their complete formula.

Our result corresponds to the natural choice of the parameters:

$$c_1 = b_0 = 2\,\mathrm{e}^{-\gamma_E},$$

$$c_2 = \frac{A_T}{Q} = \frac{\left[\left(1 - x_1^{0^2}\right)\left(1 - x_2^{0^2}\right)\right]^{1/2}}{\left(x_1^0 + x_2^0\right)}. \tag{71}$$

With these choices our expression in terms of the Bessel functions J_0 agrees with eq. (69) at the DLA. The term $-\frac{3}{2}\alpha(q_T)$, subleading in the limit $q_T^2 \to 0$, is directly present in our formula. D is the correction term given by [17,18]

$$D = \frac{1}{2\pi}\left[\tfrac{67}{6} - \tfrac{1}{2}\pi^2 - \tfrac{10}{18}n_f\right]. \tag{72}$$

In sect. 4 we shall study the quantitative importance of this non-leading correction.

Since data exist over a large range of y we report here also the expression for the cross section integrated over y. This expression has a form similar to eq. (56):

$$\frac{\mathrm{d}\sigma}{\mathrm{d}q_T^2} = N\left\{\int \frac{\mathrm{d}^2 b}{4\pi} \mathrm{e}^{-i q_T \cdot b} \exp S'(b^2, Q^2) R'(b^2, Q^2) + Y'(q_T^2, Q^2)\right\}. \tag{73}$$

The upper limit of the transverse momentum is now given by $A_T'^2 = \frac{1}{4}S(1-\tau)^2$ and consequently the form factor S becomes

$$S'(b^2,Q^2) = \int_0^{A_T'^2} dk^2 \left(J_0(bk)-1\right)\frac{\alpha_s(k^2)}{2\pi}\frac{4}{3}\left(2\frac{\ln(Q^2/k^2)}{k^2} - \frac{3}{k^2}\right). \quad (74)$$

The function R′ is given by

$$R'(b^2,Q^2) = \int dx_1 dx_2 \delta(x_1x_2-\tau) H\left(x_1,x_2,P'^2\right)$$

$$\times\left[1+\frac{\alpha_s}{2\pi}\frac{4}{3}\times\left(3\ln\frac{Q^2}{A_T'^2} - \ln^2\frac{Q^2}{A_T'^2}\right)\right] + \frac{\alpha_s}{2\pi}\frac{4}{3}\int\frac{dx_1dx_2}{x_1x_2}H\left(x_1,x_2,P'^2\right)$$

$$\times 2f_q(\tau_{12})\theta(x_1x_2-\tau) + \frac{\alpha_s}{2\pi}\frac{1}{2}\int\frac{dx_1dx_2}{x_1x_2}$$

$$\times\left[K_1\left(x_1,x_2,P'^2\right)+K_2\left(x_1,x_2,P'^2\right)\right]f_g(\tau_{12})\theta(x_1x_2-\tau), \quad (75)$$

where we define

$$\tau_{12} = \frac{Q^2}{x_1x_2S}, \qquad \rho_{12} = \frac{q_T^2}{x_1x_2S}, \qquad \rho = \frac{q_T^2}{S}, \quad (76)$$

and f_q and f_g are given in eq. (59). The corresponding finite pieces are (see eqs. (63) and (64))

$$Y_q'(q_T^2,Q^2) = \frac{4}{q_T^2}\int_\tau^1\frac{dx_1}{x_1}H\left(x_1,\frac{\tau}{x_1}\right)\ln\left[\frac{\sqrt{(x_1-\tau_+)(x_1-\tau_-)}+x_1-\tau-2\rho}{2(x_1-\tau)}\sqrt{\frac{\tau}{\tau+\rho}}\right]$$

$$-\frac{4}{S}\int\frac{dx_1dx_2}{(x_1x_2)^2}\theta(x_1x_2-\tau_+)\frac{H(x_1,x_2)}{\sqrt{(1-\tau_{12})^2-4\rho_{12}}}$$

$$+\frac{2}{q_T^2}\left\{\int\frac{dx_1dx_2}{x_1x_2}\theta(x_1x_2-\tau_+)\frac{(1+\tau_{12}^2)H(x_1,x_2)-2H(x_1,\tau/x_1)}{\sqrt{(1-\tau_{12})^2-4\rho_{12}}}\right.$$

$$\left.-\int\frac{dx_1dx_2}{x_1x_2}\theta(x_1x_2-\tau)\frac{(1+\tau_{12}^2)H(x_1,x_2)-2H(x_1,\tau/x_1)}{1-\tau_{12}}\right\}.$$

(77)

The expression for the gluonic piece is most easily expressed in terms of the quantities

$$\phi^{+}=\tfrac{1}{2}\left[1+\tau_{12}+\sqrt{(1-\tau_{12})^{2}-4\rho_{12}}\right],\qquad \phi^{-}=\frac{1}{\phi^{+}},\qquad \left(\phi^{+}\underset{q_{\mathrm{T}}^{2}\to 0}{\to}1\right).\tag{78}$$

In terms of these variables, we find

$$\begin{aligned}
Y_{\mathrm{g}}'(q_{\mathrm{T}}^{2},Q^{2})=\Bigg\{&\int\frac{\mathrm{d}x_1\,\mathrm{d}x_2}{x_1x_2}\frac{\theta(x_1x_2-\tau_{+})}{\sqrt{(1-\tau_{12})^{2}-4\rho_{12}}}\frac{1}{q_{\mathrm{T}}^{2}}\\
&\times\left[(\tau_{12}+\rho_{12})\phi^{-}-\tau_{12}-2\tau_{12}\left(\phi^{-}(\tau_{12}+\rho_{12})-\tau_{12}\right)^{2}\right]\\
&\times K_1(x_1,x_2)+\frac{1}{S}\int\frac{\mathrm{d}x_1\,\mathrm{d}x_2}{(x_1x_2)^{2}}\frac{\theta(x_1x_2-\tau_{+})}{\sqrt{(1-\tau_{12})^{2}-4\rho_{12}}}\\
&\times\left[\phi^{+}-2\tau_{12}+(\tau_{12}+\rho_{12})\phi^{-}\right]K_1(x_1,x_2)\\
&+\frac{1}{q_{\mathrm{T}}^{2}}\Bigg[\int\frac{\mathrm{d}x_1\,\mathrm{d}x_2}{x_1x_2}\theta(x_1x_2-\tau_{+})\frac{(\phi^{+}-\tau_{12})}{\sqrt{(1-\tau_{12})^{2}-4\rho_{12}}}\\
&\qquad\times\left[1-2\tau_{12}(\phi^{+}-\tau_{12})\right]K_1(x_1,x_2)\\
&\qquad-\int\frac{\mathrm{d}x_1\,\mathrm{d}x_2}{x_1x_2}\theta(x_1x_2-\tau)\left[1-2\tau_{12}(1-\tau_{12})\right]K_1(x_1,x_2)\Bigg]\Bigg\}\\
&+\{(1\leftrightarrow 2)\}.
\end{aligned}\tag{79}$$

The results for the integrals over q_T of eqs. (56) and (73) are already present in the literature [5,20]. We perform the integral over q_T of eq. (56) and obtain

$$\frac{d\sigma}{dy} = N\left\{ H(x_1^0, x_2^0, Q^2)\left[1 + \frac{\alpha_s}{2\pi}\frac{4}{3}\left(3\ln\frac{Q^2}{A_T^2} - \ln^2\frac{Q^2}{A_T^2} + I\right)\right]\right.$$
$$+ \frac{\alpha_s}{2\pi}\frac{4}{3}\left[\int_{x_1^0}^1 \frac{dz}{z} f_q(z) H(x_1^0/z, x_2^0) + \int_{x_2^0}^1 \frac{dz}{z} f_q(z) H(x_1^0, x_2^0/z)\right]$$
$$+ \frac{\alpha_s}{2\pi}\frac{1}{2}\left[\int_{x_2^0}^1 \frac{dz}{z} f_g(z) K_1(x_1^0, x_2^0/z) + \int_{x_1^0}^1 \frac{dz}{z} f_g(z) K_2(x_1^0/z; x_2^0)\right]$$
$$+ W(Q^2, y), \tag{80}$$

where

$$I = 2\int_0^{A_T^2/S} \frac{dx}{x} \ln\left[\frac{1-\sqrt{\tau+x}\,e^{y}}{1-x_1^0}\,\frac{1-\sqrt{\tau+x}\,e^{-y}}{1-x_2^0}\right]$$
$$= \mathrm{Li}_2\left(\frac{2x_1^0}{1+x_1^0}\right) + \mathrm{Li}_2\left(\frac{2x_2^0}{1+x_2^0}\right) - \tfrac{1}{3}\pi^2 + 2\ln\frac{x_1^0(1+x_2^0)}{x_1^0+x_2^0}\ln\frac{x_2^0(1+x_1^0)}{x_1^0+x_2^0}$$
$$+ \ln\frac{1+x_1^0}{1-x_1^0}\ln\frac{1+x_1^0}{2x_1^0} + \ln\frac{1+x_2^0}{1-x_2^0}\ln\frac{1+x_2^0}{2x_2^0}, \tag{81}$$

where as usual

$$\mathrm{Li}_2(x) = -\int_0^x \frac{dz}{z}\ln(1-z); \tag{82}$$

$$W(Q^2, y) = \frac{\alpha_s}{2\pi}\frac{4}{3} W_q(Q^2, y) + \frac{\alpha_s}{2\pi}\frac{1}{2} W_g(Q^2, y), \tag{83}$$

$$W_q(Q^2, y) = \int_{x_1^0}^1 \frac{dx_1}{x_1 - x_1^0}\int_{x_2^0}^1 \frac{dx_2}{x_2 - x_2^0}$$
$$\times\left[h(x_1, x_2) - h(x_1, x_2^0) - h(x_1^0, x_2) + h(x_1^0, x_2^0)\right]$$
$$- \int_{x_1^0}^1 \frac{dx_1}{x_1 - x_1^0}\ln\frac{(x_1 + x_1^0)A_T^2}{2x_1^0(x_1 - x_1^0)(1 - x_2^0)S}\left[h(x_1, x_2^0) - h(x_1^0, x_2^0)\right]$$
$$- \int_{x_2^0}^1 \frac{dx_2}{x_2 - x_2^0}\ln\frac{(x_2 + x_2^0)A_T^2}{2x_2^0(x_2 - x_2^0)(1 - x_1^0)S}\left[h(x_1^0, x_2) - h(x_1^0, x_2^0)\right]$$
$$- 2\int_{x_1^0}^1 \frac{dx_1}{x_1}\int_{x_2^0}^1 \frac{dx_2}{x_2} H(x_1, x_2)\frac{2\tau(\tau + x_1 x_2)}{(x_1 x_2^0 + x_1^0 x_2)^2}; \tag{84}$$

$$h(x_1,x_2)=\frac{2(\tau+x_1x_2)}{(x_1+x_1^0)(x_2+x_2^0)}\left[H(x_1,x_2)\left(1+\left(\frac{\tau}{x_1x_2}\right)^2\right)-2H(x_1^0,x_2^0)\right], \tag{85}$$

$$\begin{aligned}W_g(Q^2,y)=&\int_{x_2^0}^1\frac{dz}{z}\left[z^2+(1-z)^2\right]\ln\left[\frac{2(1-x_1^0)(1-z)}{(1+z)x_1^0}\right]K_1\left(x_1^0,\frac{x_2^0}{z}\right)\\&+\int_{x_1^0}^1\frac{dx_1}{(x_1-x_1^0)}\int_{x_2^0}^1 dx_2\left[K_1(x_1,x_2)-K_1(x_1^0,x_2)\right]\\&+\int_{x_1^0}^1 dx_1\int_{x_2^0}^1 dx_2\frac{2\tau(\tau+x_1x_2)}{(x_1x_2^0+x_2x_1^0)^2}\frac{x_2^0}{x_2}\left(1+\frac{x_1^{0^2}}{x_1^2}+2\frac{x_2^0}{x_2}\frac{x_1^0}{x_1}\right)\\&\times K_1(x_1,x_2)+\{(1\leftrightarrow 2)\}.\end{aligned} \tag{86}$$

After some further manipulation, the correspondence of eq. (80) with the results of refs. [5, 20] can be shown.

Lastly, we record the expression for the total cross section calculated in order α_s:

$$\begin{aligned}\sigma=&N\int\frac{dx_1}{x_1}\frac{dx_2}{x_2}H(x_1,x_2,Q^2)\left[\delta(1-\tau_{12})+\frac{\alpha_s}{2\pi}\frac{4}{3}\theta(x_1x_2-\tau)2f_q^T(\tau_{12})\right]\\&+\frac{\alpha_s}{2\pi}\frac{1}{2}\int\frac{dx_1}{x_1}\frac{dx_2}{x_2}\theta(x_1x_2-\tau)\left[K_1(x_1,x_2,Q^2)+K_2(x_1,x_2,Q^2)\right]f_g^T(\tau_{12}),\end{aligned} \tag{87}$$

where in this case the functions f^T are given by [5, 6]

$$\begin{aligned}f_q^T(z)&=\left\{\frac{3}{2}\frac{1}{(1-z)_+}+(1+z^2)\left(\frac{\ln(1-z)}{1-z}\right)_+-3-2z+\left(\tfrac{1}{2}+\tfrac{2}{3}\pi^2\right)\delta(1-z)\right\},\\f_g^T(z)&=\left\{\tfrac{1}{2}\cdot 9z^2-5z+\tfrac{3}{2}+\left(z^2+(1-z)^2\right)\ln(1-z)\right\}.\end{aligned} \tag{88}$$

4. Numerical results

Before presenting the numerical results, we discuss our treatment of the strong coupling constant and other input parameters. In performing the Fourier transform to impact parameter space, we must integrate the running coupling constant over the low-momentum region. Although this region is of little importance for the final result, for numerical reasons, however, it is convenient to introduce either a "freezing" of the coupling constant or a smearing by an intrinsic transverse momentum.

We have chosen the former procedure and fixed the running coupling constant as

$$\alpha_s(Q^2) = \left[\beta_0^{(3)} \ln \frac{Q^2 + a\Lambda^2}{\Lambda^2}\right]^{-1}, \qquad 0 \leqslant Q^2 \leqslant 4m_c^2,$$

$$\alpha_s(Q^2) = \left[\frac{1}{\alpha_s(4m_c^2)} + \beta_0^{(4)} \ln \frac{Q^2}{4m_c^2}\right]^{-1}, \qquad 4m_c^2 \leqslant Q^2 \leqslant 4m_b^2,$$

$$\alpha_s(Q^2) = \left[\frac{1}{\alpha_s(4m_b^2)} + \beta_0^{(5)} \ln \frac{Q^2}{4m_b^2}\right]^{-1}, \qquad 4m_b^2 \leqslant Q^2, \tag{89}$$

where

$$\beta_0^{(n_f)} = \frac{33 - 2n_f}{12\pi}. \tag{90}$$

This formula takes into account the change in the slope of the running coupling constant as the charm and beauty thresholds are passed, and ensures that α_s is a continuous function of Q^2. In the low-momentum region the form of the coupling constant is controlled by the parameter a ($a > 1$), which can be varied to check the sensitivity to the freezing: we checked that the results are insensitive to value of a for $q_T \gtrsim 1\ \mathrm{GeV}^2$ and $a\Lambda^2 < 1\ \mathrm{GeV}^2$.

We did not attempt to describe the smearing from the intrinsic q_T of partons inside the nucleon or that arising from initial state interactions between active and spectator quarks [21]. It is by now settled that the total production cross sections are not affected by initial state interactions [22]. However, a smearing effect can still possibly be present in the q_T distribution. The justification for neglecting these effects is that the average smearing momentum is less than $\langle q_T \rangle$ in the energy domain of interest here. As a consequence, the slight flattening of the q_T distribution from the smearing is well inside the present uncertainty on the parton component, as has been explicitly checked numerically for smearing momenta below 1 GeV.

All cross sections quoted are for the production of real ($W^+ + W^-$) or Z^0. Before comparison with data they should be multiplied by the branching ratios into the observed decay channel. The values chosen for the boson masses are

$$M_W = 83.0\ \mathrm{GeV}, \qquad M_Z = 93.8\ \mathrm{GeV}, \qquad \sin^2\theta_w = 0.217. \tag{91}$$

The sensitivity of the numerical results to the choice of quark and gluon densities in the proton (antiproton) was tested using different sets of parametrizations. We have taken the two sets of parton densities given by Duke and Owens (DO) [23] and the set proposed by Glück, Hoffmann and Reya (GHR) [24]. In fig. 3 we display the normalized q_T distribution at $y = 0$ for the charged W's for the two choices of the parton densities considered by DO. The two sets of densities are both compatible with existing data on deep inelastic scattering. The first set (DO1) has a smaller Λ ($\Lambda = 0.2$ GeV) and a narrower gluon distribution at the evolution starting point $Q_0^2 = 4$ GeV^2. The second set (DO2) has $\Lambda = 0.4$ GeV and a broader gluon distribution.

The dependence on Λ and the choice of parton densities was further studied in fig. 4 using the GHR distributions. These authors used $\Lambda = 0.4$ GeV in their analysis of the data. By comparison of fig. 3 (DO2) and fig. 4 (solid line), it turns out that the results are more sensitive to Λ rather than to the parametrization of densities. To check further the sensitivity to Λ, we also plot in fig. 4 the results for the q_T distribution of the W's obtained in the GHR case by varying Λ between 0.2 GeV and 0.6 GeV. Notice, however, that strictly speaking, the GHR distributions are only guaranteed to work for $\Lambda = 0.4$. The conclusion is that the overall uncertainty due to the choice of Λ and the form of the parton distributions is at most 25%.

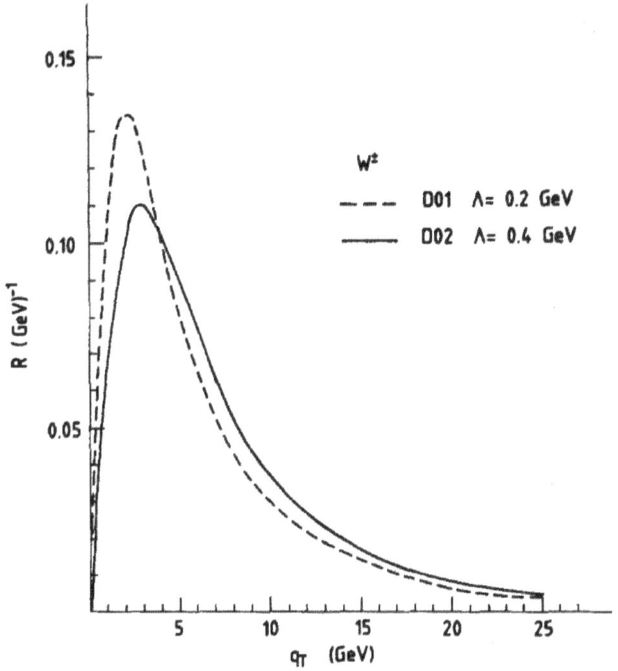

Fig. 3. The ratio $R = (d\sigma/dq_T\,dy)/(d\sigma/dy)$ at rapidity $y = 0$ as a function of q_T. The dashed line uses the Duke-Owens parametrization (set 1, $\Lambda = 0.2$ GeV) whilst the solid line shows (set 2, $\Lambda = 0.4$ GeV).

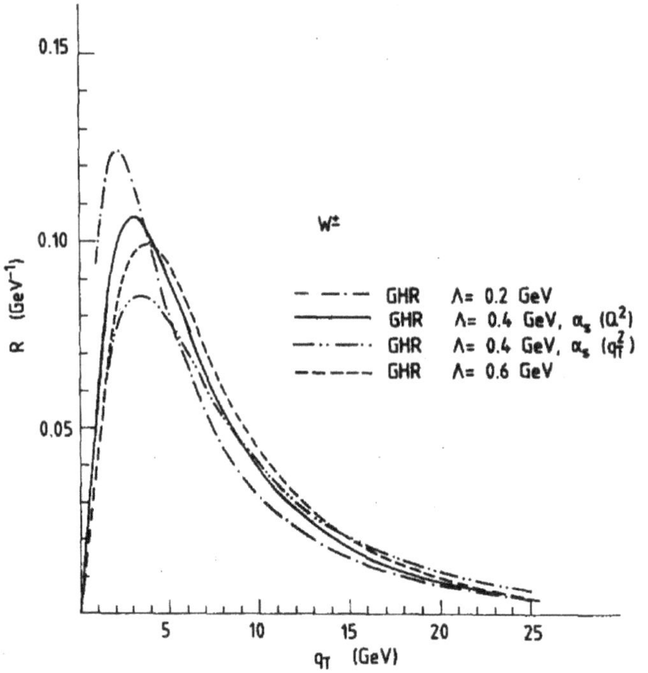

Fig. 4. The ratio $R = (d\sigma/dq_T\,dy)/(d\sigma/dy)$ at rapidity $y = 0$ using the densities of GHR. The two curves with $\Lambda = 0.4$ GeV differ by the choice of scale in terms of order α_s which is taken either as Q^2 or q_T^2 (other values of Λ correspond to the choice Q^2).

We also studied the effect of including the Kodaira-Trentadue [17] nonleading term [cf. eqs. (69), (70) and (72)]. We found that it flattens the q_T distribution by an amount that can roughly be reproduced by changing of Λ by a factor of about 1.5. This numerical check makes us confident that the effect of the neglected second-order terms, which we cannot include completely because they are not all known, is comparable with the uncertainty coming from the values of Λ and of the parton densities. All the above statements on the theoretical uncertainty hold good for the case of the Z^0 also. In the following we shall take the GHR parametrization with $\Lambda = 0.4$ GeV as a reference distribution. However, all curves are subject to the theoretical uncertainty as stated above.

In addition we recall that, for the terms of order α_s, there is a further ambiguity connected with the choice of the scale (of order Q^2) for the running coupling and the parton densities which cannot be removed without a complete knowledge of the $O(\alpha_s^2)$ terms. We have mostly taken this scale to be Q^2. An alternative would be to identify this scale with q_T^2 in the q_T distribution. The corresponding variation of the results can be seen from fig. 4 and must be kept in mind in estimating the theoretical error. It is interesting to note that the probability of producing a W with $q_T > 25$ GeV is between 3% and 6%.

In figs. 5 and 6 we plot the q_T distributions at $y = 0$ for the charged and neutral weak bosons respectively. For the case of the charged bosons, we also include the data of the UA1 and UA2 groups [25, 26] suitably normalized. Note that the normalization of our curves is determined analytically from $d\sigma/dy$ (eq. (80)). In this way we avoid a numerical normalization error due to the long tail of the q_T distribution. In fig. 7, we plot the y distribution for W production at fixed q_T in order to show that the results at $y = 0$ are actually representative of a quite large range of rapidity values. Fig. 8 displays the x_F distribution of W's plotted together with the suitably normalized histogram of 43 UA1 events.

In table 2 we report the values of the total production cross sections for $W^+ + W^-$ and Z^0 computed from eq. (87) for energies between $\sqrt{S} = 0.54$ and 2 TeV. Decay branching ratios are not included. Also reported is the ratio of the two cross sections which is less affected by theoretical uncertainties. The values shown refer to the case

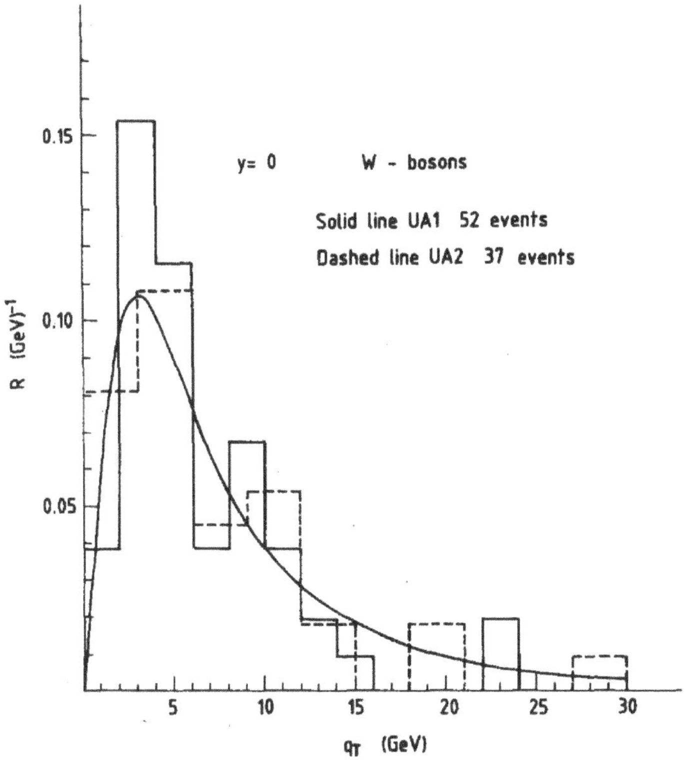

Fig. 5. The data for $W^\pm$ boson production suitably normalized and plotted against q_T. Also shown is our prediction for GHR, $\Lambda = 0.4$ GeV, $y = 0$, $\alpha_s(Q^2)$.

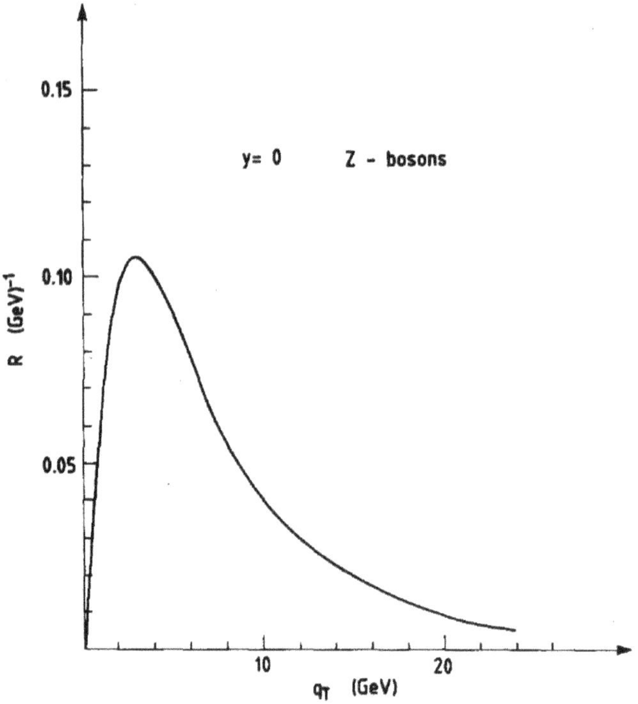

Fig. 6. R at rapidity $y = 0$ for Z^0 boson production plotted versus q_T (GHR, $\Lambda = 0.4$ GeV, $\alpha_s(Q^2)$).

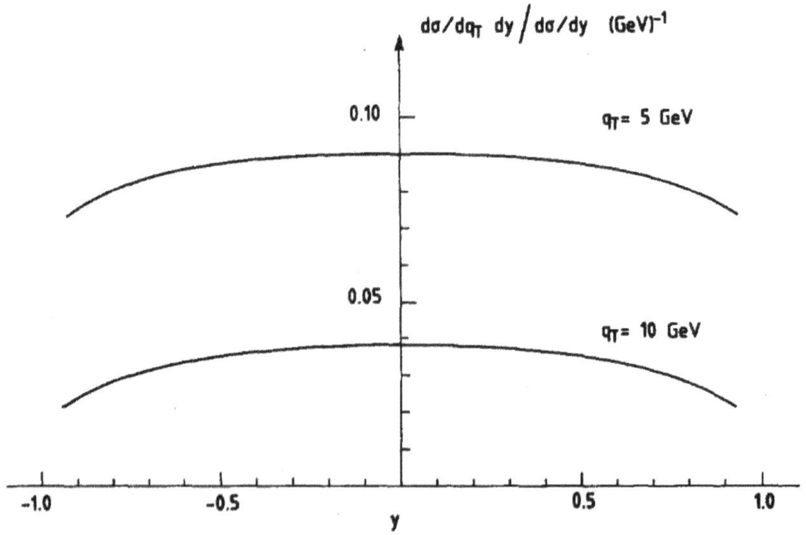

Fig. 7. The ratio $R = (d\sigma/dq_T\,dy)/d\sigma/dy$ versus y at $q_T = 5.1$ GeV; the maximum value of y is $y_{max} = 1.9$ (GHR, $\Lambda = 0.4$ GeV, $\alpha_s(Q^2)$).

where the scale Q^2 is chosen for the terms of order α_s. If so, then the $O(\alpha_s)$ terms included in these results represent about 30% of the total. However, it could be argued that since the bulk of the cross sections arises at relatively small values of q_T^2, the relevant scale could be smaller, in the range $\langle q_T^2 \rangle - Q^2$. This would imply larger values for the total cross sections. We therefore include this theoretical uncertainty in the errors which we quote in the following.

At $\sqrt{S} = 540$ GeV, we find that the total cross sections are

$$\sigma^{W^+ + W^-} = 4.2 \pm {}^{1.3}_{0.6} \text{ nb}, \quad \sigma^{Z^0} = 1.3 \pm {}^{0.4}_{0.2} \text{ nb}. \tag{92}$$

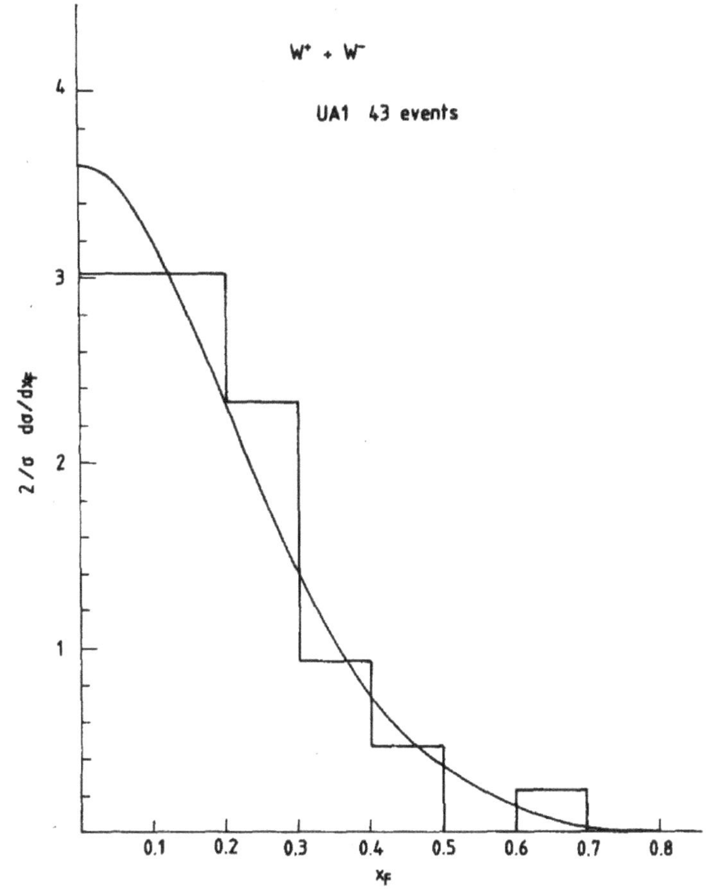

Fig. 8. The prediction for $W^\pm$ production versus x_F. The plot shows $(2/\sigma_{TOT})(d\sigma/dx_F)$ versus x_F. The data are from the UA1 Collaboration (GHR, $\Lambda = 0.4$ GeV, $\alpha_s(Q^2)$).

TABLE 2
Values (in nb) of the total cross sections for $W^{\pm}$ and Z^0 production

$\sqrt{S}$ (GeV)	$W^+ + W^-$ GHR	$W^+ + W^-$ DO1	$W^+ + W^-$ DO2	Z^0 GHR	Z^0 DO1	Z^0 DO2	$\frac{\sigma(W^+ + W^-)}{\sigma(Z^0)}$ GHR	$\frac{\sigma(W^+ + W^-)}{\sigma(Z^0)}$ DO1	$\frac{\sigma(W^+ + W^-)}{\sigma(Z^0)}$ DO2
540	4.2	4.3	4.1	1.3	1.3	1.2	3.1	3.4	3.5
700	6.2	6.3	6.1	2.0	1.9	1.8	3.1	3.3	3.4
1000	9.5	9.5	9.6	3.1	3.0	2.9	3.1	3.2	3.3
1300	12.5	12.5	12.9	4.0	3.9	3.9	3.1	3.2	3.3
1600	15.5	15.6	16.5	5.0	4.8	5.0	3.1	3.2	3.3

The ratio of the two cross sections is less affected by theoretical errors:

$$\frac{\sigma^{W^+ + W^-}}{\sigma^{Z^0}} = 3.3 \pm 0.2\,, \tag{93}$$

and is important in order to obtain Γ_W/Γ_Z from experiment [31]. Multiplying by the branching ratio into electrons:

$$B(W \to e\nu) = 0.089\,, \qquad B(Z^0 \to e^+e^-) = 0.032\,, \tag{94}$$

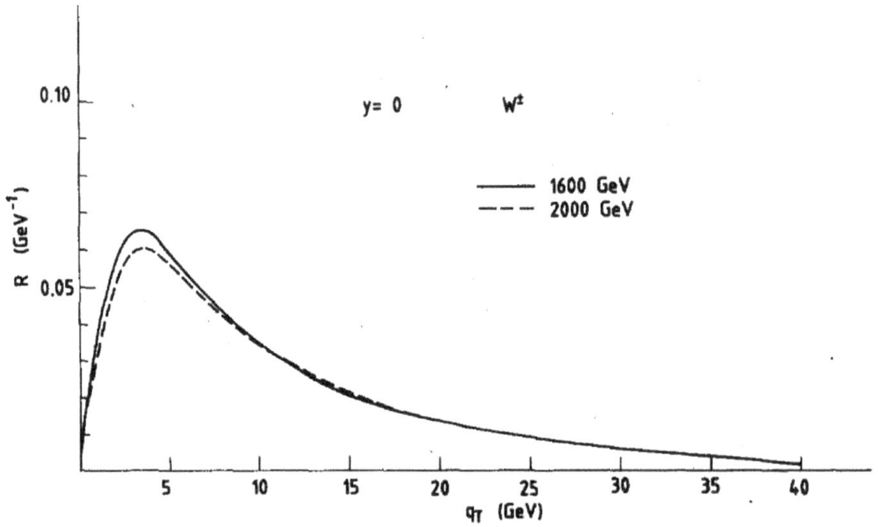

Fig. 9. Prediction for the W distribution in transverse momentum at the tevatron energies: $\sqrt{S} = 1600$ GeV, solid line; $\sqrt{S} = 2000$ GeV, dashed line. Similar curves can be drawn for the Z^0 distribution (GHR, $\Lambda = 0.4$ GeV, $\alpha_s(Q^2)$).

which are the values obtained for the top quark mass $m_t = 40$ GeV and $\alpha_s/\pi = 0.04$, we find that the product of the cross section and decay branching ratio is

$$(\sigma B)^{W^\pm \to e^\pm} = 370 \pm ^{110}_{60} \text{ pb}, \quad (\sigma B)^{Z^0 + e^+e^-} = 42 \pm ^{12}_{6} \text{ pb}. \tag{95}$$

The results in eq. (94) are in perfect agreement with those of ref. [27], where the total cross sections are also computed from eq. (87) with α_s at the scale Q^2. On the other hand, we do not reproduce the results of ref. [28], especially at higher energies. A somewhat larger result has been found in ref. [29], where, however, the total cross section is reconstructed from the lepton p_T distribution by a numerical integration. Finally, a value for the Z^0 production cross section of about 55 pb can be derived from the work of ref. [30]. The corresponding experimental results are [25, 26]

$$\text{UA1:} \quad (\sigma B)^{W^\pm} = 530 \pm 80 \pm 90 \text{ pb}, \qquad (\sigma B)^{Z^0} = 71 \pm 24 \pm 13 \text{ pb},$$

$$\text{UA2:} \quad (\sigma B)^{W^\pm} = 530 \pm 100 \pm 100 \text{ pb}, \quad (\sigma B)^{Z^0} = 110 \pm 40 \pm 20 \text{ pb}.$$

Finally, in fig. 9 we give the distribution in the transverse momentum of the W's at the tevatron energies $\sqrt{S} = 1.6, 2$ TeV.

We thank G. Pancheri, G. Parisi, D.E. Soper and W.J. Stirling for useful discussions.

References

[1] UA1 Collaboration, G. Arnison et al., Phys. Lett. 122B (1983) 103; 126B (1983) 398
[2] UA2 Collaboration, G. Banner et al., Phys. Lett. 122B (1983) 676; 129B (1983) 130
[3] S. Drell and T.M. Yan, Phys. Rev. 25 (1970) 316; Ann. of Phys. 66 (1971) 578
[4] G. Altarelli, Phys. Reports 81 (1982) 1
[5] G. Altarelli, R.K. Ellis and G. Martinelli, Nucl. Phys. B157 (1979) 461
[6] J. Kubar-André and F.E. Paige, Phys. Rev. D19 (1979) 221
[7] G. Parisi, Phys. Lett. 90B (1980) 295;
G. Curci and M. Greco, Phys. Lett. 92B (1980) 175
[8] G. Altarelli, G. Parisi and R. Petronzio, Phys. Lett. 76B (1978) 351, 356
[9] H. Fritzsch and P. Minkowski, Phys. Lett. 74B (1978) 384;
K. Kajantie and R. Raitio, Nucl. Phys. B139 (1978) 72;
F. Halzen and P.M. Scott, Phys. Rev. D18 (1978) 3378;
P. Aurenche and J. Lindfors, Nucl. Phys. B185 (1981) 274;
D. Soper, Phys. Rev. Lett. 38 (1977) 461
[10] Yu.L. Dokshitzer, D.I. Dyakonov and S.I. Troyan, Phys. Lett. 78B (1978) 290; Phys. Reports 58 (1980) 269
[11] G. Parisi and R. Petronzio, Nucl. Phys. B154 (1979) 427;
G. Curci, M. Greco and Y. Srivastava, Phys. Rev. Lett. 43 (1979) 434; Nucl. Phys. B159 (1979) 451
[12] J.C. Collins and D.E. Soper, Nucl. Phys. B193 (1981) 381; B194 (1982) 445; B197 (1982) 446
[13] S.-C. Chao and D.E. Soper, Nucl. Phys. B214 (1983) 405;
S.-C. Chao, D.E. Soper and J.C. Collins, Nucl. Phys. B214 (1983) 513

[14] F. Halzen, A.D. Martin and D.M. Scott, Phys. Rev. D25 (1982) 754;
P. Chiappetta and M. Greco, Nucl. Phys. B221 (1983) 269; Phys. Lett. 135 (1984) 187;
P. Aurenche and R. Kinnunen, Phys. Lett. 135B (1984) 493
A. Nakamura, G. Pancheri and Y. Srivastava, Z. Phys. 21 (1984) 243
[15] F. Halzen, A.D. Martin and D.M. Scott, Phys. Lett. 12B (1982) 160
[16] R.K. Ellis, G. Martinelli and R. Petronzio, Nucl. Phys. B211 (1983) 106
[17] J. Kodaira and L. Trentadue, Phys. Lett. 112B (1982) 66; 123B (1983) 335
[18] C.T.H. Davies and W.J. Stirling, CERN preprint TH.3853 (1984)
[19] S.D. Ellis, N. Fleishorn and W.J. Stirling, Phys. Rev. D24 (1981) 1386
[20] J. Kubar-André, M. Le Bellac, J.L. Mennier and G. Plaut, Nucl. Phys. B175 (1980) 251
[21] G.T. Badwin, S.J. Brodsky and G.P. Lepage, Phys. Rev. Lett. 47 (1981) 1799
[22] W.W. Lindsay, D.A. Ross and C.T. Sachrajda, Nucl. Phys. B214 (1983) 61
[23] D.W. Duke and J.F. Owens, Florida State Univ. preprint FSU-HEP 831115 (1984)
[24] M. Glück, E. Hoffmann and E. Reya, Z. Phys. C13 (1982) 119
[25] UA1 Collaboration, presented by C. Rubbia, Proc. 4th $p\bar{p}$ Workshop, Bern (1984)
[26] UA2 Collaboration, presented by J. Schacher, Proc. 4th $p\bar{p}$ Workshop, Bern (1984)
[27] P. Chiappetta and M. Perrottet, Marseille preprint CPT-83P-1525 (1983)
[28] F. Paige, Proc. Topical Workshop, Production of new particles at superhigh pp collisions, ed. V. Barger et al. (Univ. of Wisconsin, Madison, 1979)
[29] P. Minkowski, Bern preprint BUTP-83/22 (1984)
[30] B. Humpert and W.L. van Neerven, Phys. Lett. 93B (1980) 456;
B. Humpert, private communication
[31] N. Cabibbo, Proc. 3rd $\bar{p}p$ Workshop, Rome, ed. C. Bacci and G. Salvini, 1983

www.ingramcontent.com/pod-product-compliance
Lightning Source LLC
Chambersburg PA
CBHW081517190326
41458CB00015B/5394
9789811213892